Sex, Color, and Mate Choice in Guppies

MONOGRAPHS IN
BEHAVIOR AND ECOLOGY

Edited by John R. Krebs and
Tim Clutton-Brock

Five New World Primates: A Study in Comparative Ecology, *by John Terborgh*

Reproductive Decisions: An Economic Analysis of Gelada Baboon Social Strategies, *by R.I.M. Dunbar*

Honeybee Ecology: A Study of Adaptation in Social Life, *by Thomas D. Seeley*

Foraging Theory, *by David W. Stephens and John R. Krebs*

Helping and Communal Breeding in Birds: Ecology and Evolution, *by Jerram L. Brown*

Dynamic Modeling in Behavioral Ecology, *by Marc Mangel and Colin W. Clark*

The Biology of the Naked Mole-Rat, *edited by Paul W. Sherman, Jennifer U. M. Jarvis, and Richard D. Alexander*

The Evolution of Parental Care, *by T. H. Clutton-Brock*

The Ostrich Communal Nesting System, *by Brian C.R. Bertram*

Parasitoids: Behavioral and Evolutionary Ecology, *by H.C.J. Godfray*

Sexual Selection, *by Malte Andersson*

Polygyny and Sexual Selection in Red-winged Blackbirds, *by William A. Searcy and Ken Yasukawa*

Leks, *by Jacob Höglund and Rauno V. Alatalo*

Social Evolution in Ants, *by Andrew F. G. Bourke and Nigel R. Franks*

Female Control: Sexual Selection by Cryptic Female Choice, by William G. Eberhard

Sex, Color, and Mate Choice in Guppies, *by Anne E. Houde*

Sex, Color, and Mate Choice in Guppies

ANNE E. HOUDE

Princeton University Press
Princeton, New Jersey

Published by Princeton University Press,
41 William Street,
Princeton, New Jersey 08540
In the United Kingdom: Princeton University Press,
Chichester, West Sussex

Library of Congress Cataloging-in-Publication Data

Houde, Anne E., 1959–
Sex, color, and mate choice in guppies / Anne E. Houde.
p. cm. — (Monographs in behavior and ecology)
Includes bibliographical references (p.) and index.
ISBN 0-691-02790-0 (cloth : alk. paper) —
ISBN 0-691-02789-7 (Pbk. : alk. paper)
1. Guppies—Behavior. 2. Sexual behavior in animals. 3. Sexual selection in animals. I. Title. II. Series.
QL638.P73H67 1997
591'.667—dc21 96-49200

This book has been composed in Times Roman

Princeton University Press books are
printed on acid-free paper and meet the guidelines
for permanence and durability of the Committee
on Production Guidelines for Book Longevity
of the Council on Library Resources

Printed in the United States of America by Princeton Academic Press

10 9 8 7 6 5 4 3 2 1

10 9 8 7 6 5 4 3 2 1
(Pbk.)

To Jonathan and Ari

Contents

Acknowledgments

The work on guppies described in this book might not have come about except for the trail-breaking work done by the early pioneers who brought the potential of the species to the attention of the world. These include Ö. Winge, who was among the first to embark on scientific studies of guppies; Eugenie Clark and Lester Aronson, who uncovered basic information about the reproductive biology and behavior of guppies, and G. P. Baerends and coworkers, who put together the ethological details of guppy sexual behavior. These earliest studies paved the way for fieldwork on the evolutionary ecology of guppies by Caryl and Edna Haskins and Robin Liley. These researchers are responsible for bringing a modern evolutionary approach to the study of sexual selection in guppies and providing a basic understanding of the species in nature. The Haskinses were responsible for bringing the extraordinary geographic variation in guppies to the attention of biologists. Ironically, the studies by Liley and the Haskinses addressed questions about the role of sexual selection in speciation and reproductive isolation that even now remain unanswered and warrant further research efforts. I thank all of these people for their efforts in laying the groundwork for later work on sexual selection in guppies.

Once these foundations were laid, research on the behavioral and evolutionary ecology of guppies increased rapidly. The people whose work contributed to the field, inspired my own studies, and motivated the writing of this book have become too numerous to detail here. Their contributions are described in the chapters that follow. I must recognize the contribution of one person, however, whose elegant studies were responsible for sparking my interest in sexual selection in guppies, and who was personally responsible for encouraging me throughout my career and throughout the writing of this book. John Endler began where the Haskinses left off and provided clear demonstrations of the effect of predators on the evolution of guppies. More importantly, John has never been content with his latest area of study, but is always taking new directions with his work while also leading the research efforts of others in new directions. Thus, once he had obtained evidence on the effects of predators on the evolution of guppy color patterns, he moved on to an interest in the role of sexual selection, at the same time kindling my own interest. He then began examining the physical and sensory basis for sexual communication, again fostering work by new researchers. I therefore extend my thanks to John for both his personal encouragement of my work and his leadership in the field.

I would also like to thank numerous other colleagues for the encouragement, feedback, and discussions they have provided and for keeping me informed of their new results during the writing of this book. The community of researchers who work on guppies and sexual selection is an example of what science as a social endeavor ought to be. I am especially grateful to Alex Basolo, Gerry Borgia, Felix Breden, Rob Brooks, Lee Dugatkin, Jean-Guy Godin, Anson Gong, Astrid Kodric-Brown, Robin Liley, Kevin Long, Annarie Lyles, Anne Magurran, Paul Nicoletto, Shawn Nordell, Margaret Ptacek, John Reynolds, David Reznick, Helen Rodd, Nilla Rosenqvist, Pat Ross, Dan Rubenstein, Victor Rush, and Tish Sheridan for their help, encouragement, and inspiration. I also thank the students who have assisted with my work on guppies over the years. These include Becky Bordeau, Bradd Bridges, Licia Buono, Nancy Carl, Christine Case, Pat Cassidy, Walid Chaloub, Milda Cimermanis, Debbie Coplin, Carl Deetz, Kirsten Findlay, Melissa Hankes, Natasha Holbert, Julie King, Lelsie Malmgren, Brigitte Nguyen, Peter Nicinski, Carolyn Renshaw, John Rumbaugh, Helen Shade, Jason Stanley, Joe Torio, and James Williams.

The help of several people has been invaluable to me while I was writing this book, and I am especially indebted to them. John Endler, Anne Magurran, John Reynolds, David Reznick, and Helen Rodd read and commented on the manuscript, suggested improvements, set me straight on numerous points, and gently pointed out muddy patches in my thinking and writing. I am responsible for all remaining deficiencies. I am grateful for the hard work of Peter Nicinski, who among other things, unfailingly checked and rechecked the references, performed electronic searches, and photographed guppies, for ten long weeks. Finally, words are inadequate to express my gratitude to Jonathan Weiland, not only for creating most of the figures and spending time at night and on weekends on tedious computer work for the book, but most importantly for his constant support, understanding, and encouragement.

I would like to thank Annarie Lyles for providing a photograph, and Bailey Donnally, Peter Nicinski, and Jonathan Weiland for photographing guppies. I am grateful to the National Science Foundation for financial support.

Sex, Color, and Mate Choice in Guppies

1 Introduction: The Guppy as a Model System

Since the early part of this century, the guppy (*Poecilia reticulata*) has been the subject of scientific inquiry in several disciplines, and at the same time has become dear to the hearts of aquarists of all ages. Guppies are best known for their conspicuous color patterns and for the incessant courtship of females by males. These traits have been the focus of studies of sexual selection in this species and are the subject of this book. Some of the earliest work focused on the genetics of color pattern inheritance. Subsequently, the reproductive biology, sexual behavior, and ecological genetics of guppies attracted attention. Because of their easy availability, guppies are a favorite choice for toxicology studies (unfortunately for them!). Most recently, and currently, guppies have proved to be ideal subjects for experiments in behavioral ecology. In particular, the nature of the guppy mating system makes guppies especially amenable to studies of sexual selection and mate choice. There is now a substantial body of literature on sexual selection and mate choice in guppies, and new experiments continue to be devised. Beginning at a time when developments in sexual selection theory outstripped empirical results, work on guppy sexual behavior in recent years has provided much needed information on how sexual selection may operate in natural populations.

The goals of this monograph are to summarize and synthesize the work on sexual selection and mate choice in guppies, to relate the empirical findings on guppies to current themes in sexual selection theory and empirical work with other species, and to suggest profitable directions for further work. This introductory chapter introduces the evolution, genetics, and ecology of the species, describes the geographic variation that has made the guppy system such a productive natural laboratory for evolutionary studies, and provides a brief overview of the issues in sexual selection that have been studied with the guppy system. An introduction to the reproductive biology and sexual behavior of guppies is presented in chapter 2.

1.1 Evolution, Genetics, and Ecology of Guppies

Current interest in sexual selection in guppies and in the utility of the species as a model system stems from a number of key observations about the evolutionary ecology and genetics of guppies, especially with refer-

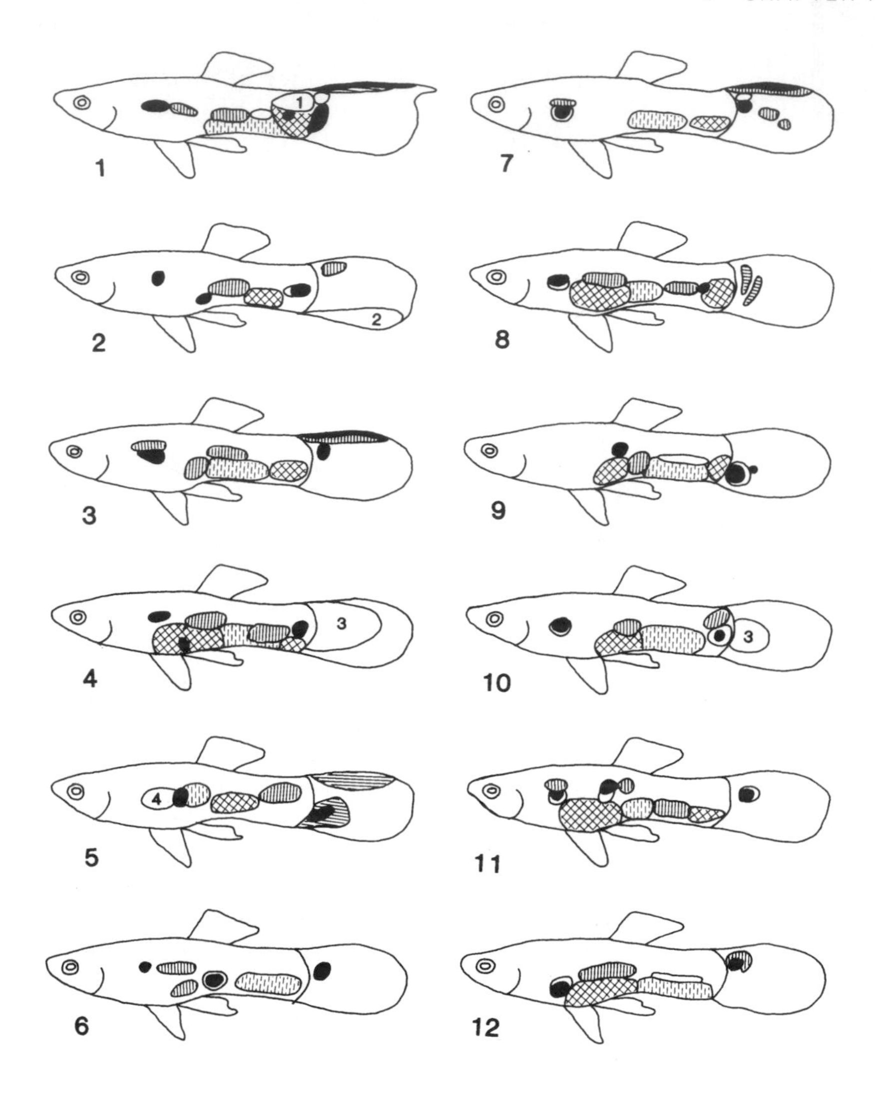

Figure 1.1 Diagrams of representative color patterns of male guppies. These fish were from a stock originally collected from the upper Aripo River, Trinidad. Shading indicates colors of spots—black: black; unshaded: white; vertical hatching: orange; horizontal hatching: yellow; cross-hatching: green; broken hatching: blue/violet. Numbered spots: 1, cream; 2, brown; 3, black and yellow speckling; 4, brick red.

ence to color patterns. The color patterns of guppies are complex and often conspicuous combinations of black, white, red-orange, yellow, green, iridescent, and other spots, speckles, and lines. Color patterns are expressed only in males and show a great deal of heritable variation among individu-

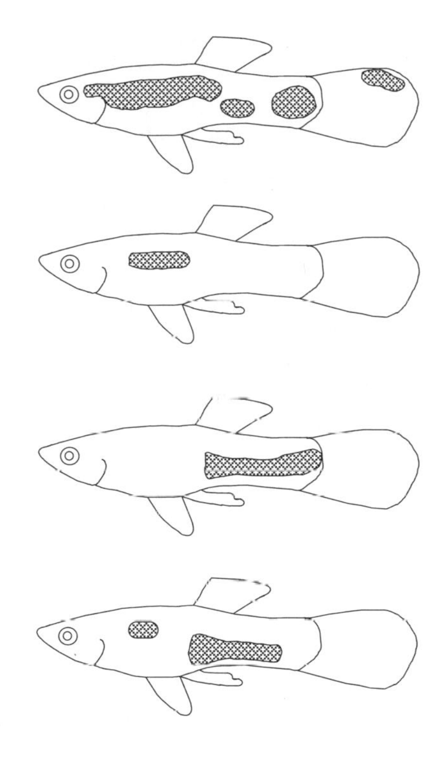

Figure 1.2 Illustrations of variation in orange-red carotenoid spots (shaded) of male guppies from the Paria River, Trinidad.

als. In a collection of as many as thirty or more males, each individual is likely to have a unique pattern (figs. 1.1, 1.2). Studies early in this century by Winge examined the inheritance of a number of color pattern morphs and demonstrated X-and Y-linked inheritance of entire patterns. These observations brought the remarkable polymorphism of guppy color patterns to the attention of biologists and motivated work on the ecological genetics of color patterns. It is ironic, however, that we still do not understand how the polymorphism of color patterns originates and is maintained in guppy populations.

Beginning in the 1940s, a series of ground-breaking field and laboratory studies by Caryl and Edna Haskins and coworkers examined the population biology and evolutionary ecology of guppies and related species. They showed that variation in color patterns included not only within-population polymorphism but also differences among populations that seemed to be related to differences in predation regime.

The role of predation regime in geographic variation in color patterns was then demonstrated elegantly and rigorously by John Endler in the 1970s. Predation has since been shown to shape the evolution of many characters in guppies besides color patterns, including life history traits, courtship behavior, and antipredator behavior.

Systematics and Distribution

Guppies belong to the cyprinodont family Poeciliidae, the livebearers. Meffe and Snelson (1989a) have gathered extensive information on the biology of this family, including much comparative information that will not be repeated here. See Rosen and Bailey (1963) and Parenti and Rauchenberger (1989) for information on systematics of the family. All poeciliids have internal fertilization and give birth to live young (see chapter 2).

Besides guppies, a number of other poeciliid species are also well known in the field of behavior and evolution. These include sailfin mollies (*Poecilia latipinna*), known for polymorphism and plasticity in male size, rate of maturation, and courtship strategies (e.g., Woodhead and Armstrong 1985; Farr et al. 1986; Travis and Woodward 1989; Trexler and Travis 1990; Travis and Trexler 1990; Travis 1994); swordtails and platyfish (*Xiphophorus spp.*), known for intra- and interspecific variation in sexual ornaments and courtship behavior (Ryan and Wagner 1987; Ryan and Causey 1989; Zimmerer and Kallman 1989; Basolo 1990a, 1995a,b; Meyer et al. 1994) and for unusual patterns of chromosomal sex determination (e.g., Kallman 1983); and various species of *Poeciliopsis*, known for unusual forms of sexual and asexual reproduction (e.g., Vrijenhoek 1989).

Guppies and closely related *Poecilia* species are all small tropical livebearers. Other *Poecilia* species (*P. vivipara*, *P. parae* and *P. picta*) are sympatric with guppies in some places but they do not appear to hybridize successfully (Haskins and Haskins 1949, 1950; Liley 1966). Guppies are notable among their close relatives for having the greatest elaboration of color patterns and the greatest degree of sexual dimorphism and polymorphism in color pattern. Most other related species are monomorphic in color pattern or have only a few color pattern morphs. Only males express color patterns, and female guppies often grow much larger than males.

Guppies are native to streams and rivers of Trinidad and Tobago and adjacent parts of South America. They have now been introduced all over the world, especially for mosquito control in ponds in tropical countries. Most of the fieldwork on guppies has been conducted in streams of the Northern Range of Trinidad (fig. 1.3), although Winemiller et al. (1990) have recently examined variation in Venezuelan populations of guppies. All of the river and stream names given in this and subsequent chapters refer to guppy localities in Trinidad.

Genetics

The clearly visible polymorphism of male color patterns made guppies a good subject for early genetic studies. Guppies have typical X-Y sex determination, with heterogametic males. Winge's work early in the century

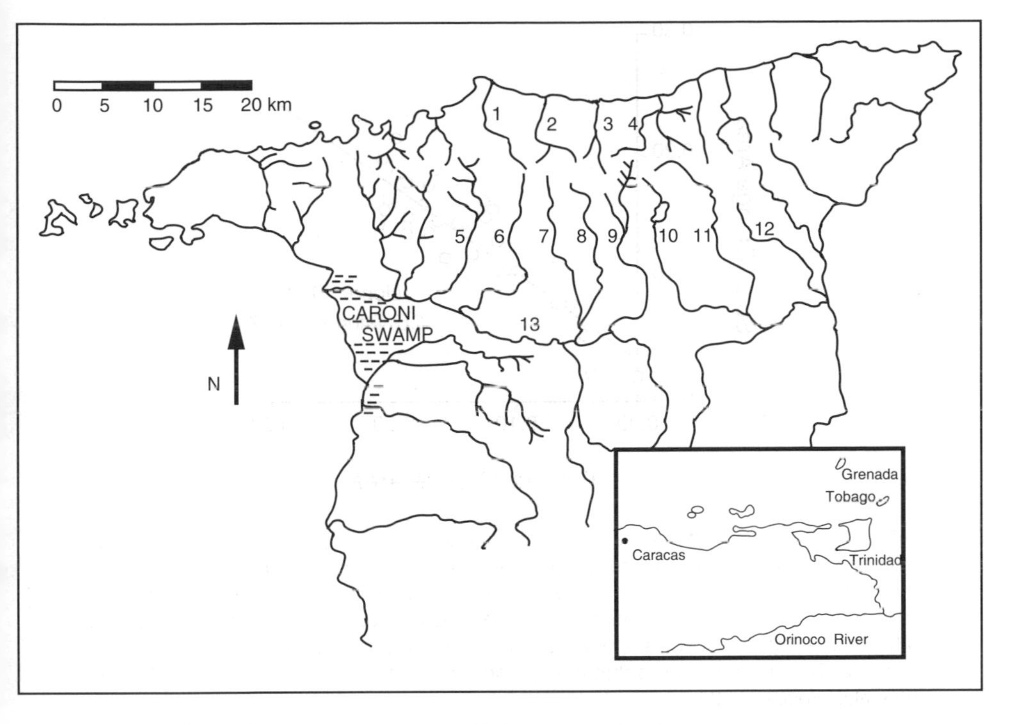

Figure 1.3 Map of the Northern Range of Trinidad showing locations of guppy research and collections. *Inset*: location map showing Trinidad's position in the southeast Caribbean. Some of the principal rivers in which guppies have been studied: 1, Yarra; 2, Marianne; 3, Paria; 4, Madamas; 5, Tacarigua; 6, Arouca; 7, Arima; 8, Guanapo; 9, Aripo; 10, Quare; 11, Oropuche; 12, Rio Grande; 13, Caroni.

(Winge 1922a,b, 1923, 1927; Winge and Ditlevsen 1947) demonstrated that the color patterns of the different morphs that had been collected and maintained in the laboratory were generally inherited as a unit and show either X or, more commonly, Y linkage. In general, male guppies usually inherit the color pattern of their fathers intact, except possibly for variation in the pattern on the tail. The Y linkage of guppy color patterns led Fisher (1930a) to argue that characters which are advantageous to males and disadvantageous to females should accumulate on sex chromosomes. The high level of polymorphism in X- and Y-linked color patterns in guppies is more puzzling and needs further study. According to models by Clark (1987), for example, such sex-linked polymorphism can be maintained by natural selection only in unusual genetic systems.

Winge studied the inheritance of a considerable number of color patterns, which he termed morphs, but, in fact, the number of different color patterns in natural populations is even larger. There are at least forty color pattern loci that can be recognized with some frequency (Endler 1978, ref-

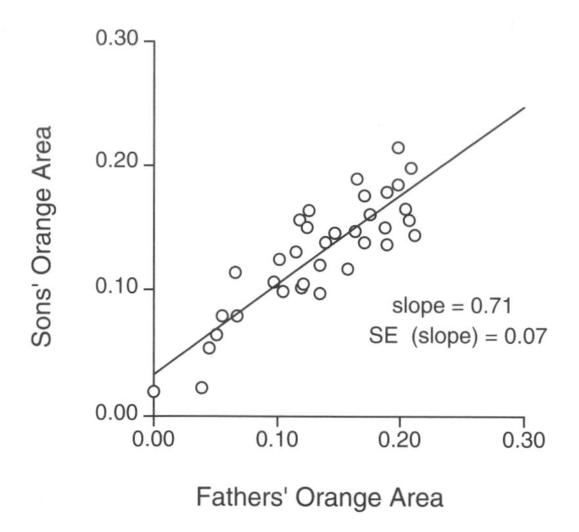

Figure 1.4 Regression of the relative area of orange in male guppies on mean range area of their sons. Relative orange area is the area of orange spots divided by the total area of the body, not including fins. Fathers were laboratory reared, and one to four sons were measured from each family. Notice that the regression slope is significantly greater than 0.5. For an autosomal trait, a slope of 0.5 would indicate a heritability of 1.0, the maximum possible. This result suggests that the orange area trait may have substantial Y linkage. (See also Houde 1992.)

erences therein, and pers. comm.). Several different loci can be expressed in a single color pattern, resulting in a huge number of possible color pattern genotypes.

Much of the early literature on guppies refers to particular color pattern "strains" in which individuals have identical color patterns or share specific color pattern elements. Because they have a well-defined genetic basis, these different strains are more analogous to the various mutants of *Drosophila melanogaster*, rather than to geographical races or artificially bred varieties of domesticated animals. Several of the strains are laboratory stocks descended from particular color pattern types studied originally by Winge (e.g., *maculatus*, *armatus*, *pauper*) and were maintained in several laboratories, notably that of Caryl Haskins, for many years. Other strains have been obtained more recently either from the wild or as mutations that have arisen in captivity. In many of the named strains (e.g., *pauper*, *armatus*, *maculatus*, etc.) studied by the Haskinses (1949, 1950, 1951, 1954, 1961, 1970) and Farr (1976, 1977, 1980a,b, 1981, 1983), the color patterns are Y-linked and the strains are probably descended from a single ancestral male. Y-linked color patterns can be traced easily through male lines, since they are inherited in an essentially clonal fashion.

While some of the known sex-linked color pattern elements are easily identified as discrete characters that have measurable frequencies in populations, color pattern characters can also be measured quantitatively, for example, in terms of the number, position, and size of particular color spots. Using standard quantitative genetic techniques (father-son regressions, half-sib analyses, selection experiments), I found that one such quantitative measure, the relative area of orange pigment in a color pattern, has high heritability (at least 0.70) and shows evidence of Y linkage (Houde 1992; see figs. 1.2, 1.4). This is not surprising given the known sex-linked inheritance of elements that contribute to the quantitative characteristics of the color pattern as a whole. This high heritability indicates that color patterns have the potential to evolve rapidly in response to natural and sexual selection in natural populations (see chapter 6 for discussion of artificial selection experiments and further discussion of the implications of high heritability for sexual selection models).

Some information is has been obtained on the inheritance of characters other than color patterns. Farr (1983; see also Farr and Peters 1984) examined the genetic basis for differences in courtship behavior between laboratory strains and inbred lines and showed an effect of the Y chromosome in some cases. Body size seems to have high heritability in guppies and shows correlations with other related life history traits (e.g., Reynolds and Gross 1992; Reznick et al., submitted). The inheritance of body size is so clear-cut, despite considerable environmental effects, that some have speculated that body size may be determined in a way similar to the Y linked P-allele system of *Xiphophorus* species (Kallman and Borkoski 1978; Kallman 1989).

ECOLOGY

Typical guppy habitats in Trinidad are clear, swiftly flowing, relatively sterile streams in mountain forest areas (see fig. 1.5), but guppies also occur in larger, more turbid lowland streams. Even headwater streams can flood and become murky during persistent rains. There is generally little or no underwater or emergent vegetation in these streams, so the visual background for guppies and their predators is rocks, multicolored gravel, and leaf litter. Disturbed sites may have more siltation and vegetation growth. Most guppies are found in freshwater; however, they have good tolerance for salinity in the laboratory and a few populations live in brackish waters. In natural habitats, guppies prefer relatively still pools in small streams and the edges and backwaters of rivers. Only large females and a few males can be found in deeper, more swiftly moving parts of streams (Haskins et al. 1961; Mattingly and Butler 1994). In the laboratory, guppies show a gen-

Figure 1.5 A typical guppy stream in Trinidad (low-predation headwater stream).

eral tendency to prefer shallow over deeper microhabitats (Noltie and Johanson 1986). These microhabitat preferences may depend on rearing conditions and the presence of predators and other guppies. See Meffe and Snelson (1989b) for further information about habitats and ecology of poeciliid fishes.

Reproduction by guppies occurs year round but may be limited by lack of food during the wet season (Reznick 1989). There have been a few laboratory studies of population dynamics using guppies as a model system (see Dahlgren 1979; Barlow 1992 and references therein), but relatively little is known about population dynamics in nature. Aspects of population structure such as sex ratios and size structure have been studied by Haskins et al. (1961), Seghers (1973), and Rodd (1994).

Guppies are omnivorous, opportunistic feeders. They consume insect larvae and other invertebrates, algae and benthic detritus (Dussault and Kramer 1981), their own young, and the eggs and young of *Rivulus hartii* (D. Fraser, pers. comm.), a cyprinodontid fish that occurs sympatrically and is also a predator of guppies. There is relatively little detailed information on diet choice and feeding behavior of guppies. Dussault and Kramer (1981) found that algae made up about 50% of the diet of wild Trinidad guppies sampled from two streams. In laboratory experiments, they confirmed that guppies can grow on a diet of green algae alone, but grow faster when the diet includes animal material. Male and female guppies differ in feeding behavior in a number of respects. Males are able to ingest more food per peck than females when feeding on algae, and spend much less time foraging (Dussault and Kramer 1981; Magurran and Seghers 1994c). In single-sex groups, males move between feeding sites more frequently than do females, and females appear to rely more on previous experience in choosing a feeding site than do males (Abrahams 1989). These differences probably reflect higher priority placed on searching for mates and courtship by males. The trade-off between feeding and courting is discussed in more detail in chapter 4.

Predation has a profound influence on many aspects of the biology of guppies. In particular, for our interests, predation imposes costs on the expression of male color patterns and sexual displays and may also impose costs of mate choice on females. The evidence that selective predation affects the evolution of color patterns is especially strong for guppies (Haskins et al. 1961; Endler 1978, 1980, 1983). Mortality costs of conspicuous secondary sexual characters have been documented in other fish species such as threespine sticklebacks *Gasterosteus aculeatus* (Moodie 1972), African killifish *Notobranchius guentheri* (Haas 1976), and also in birds (Promislow et al. 1992), crickets *Gryllus integer* (Cade 1975), and frogs *Physalaemus pustulosus* (Tuttle and Ryan 1982; Ryan et al. 1982). See Magnhagen (1991) and Andersson (1994) for additional examples.

Guppies show both short-term behavioral responses to predators (see chapters 4 and 5 for effects on male courtship and female mate choice), demographic responses (Rodd 1994), and long-term evolutionary responses. The well-documented patterns of geographic variation in color patterns and other traits among guppy populations (see sec. 1.2) stem to a large degree from differences in the assemblages of predators that occur in different streams. Some headwater streams in Trinidad are relatively free of predatory fishes. In these, the only predators are *Rivulus hartii*, a cyprinodontid, and sometimes *Macrobrachium crenulatum* (a freshwater prawn). These predators are probably not very dangerous to guppies (Endler 1978, 1983). More dangerous predators occur in the main trunks of streams and in lowland rivers. *Crenicichla alta*, the pike cichlid, is one of the most dangerous diurnal predators of guppies and probably has the greatest effects on the evolution of a variety of characters. Many of the comparisons of guppy populations refer to high predation localities as "*Crenicichla* localities" and to low predation localities as "*Rivulus* localities." Table 1.1 presents a list of some of the fish that are known to prey on guppies in Trinidad and Venezuela. Within Trinidad, different assemblages of piscivorous fishes are found in streams flowing north from the mountains of the Northern Range into the Caribbean and those flowing south into the Atlantic or the Gulf of Paria. Other likely predators on guppies include kingfishers, herons, bats, beetles, hemiptera, and spiders (Seghers 1973; Endler 1978, 1983; Winemiller et al. 1990), but the impact of these species is not well understood.

The different predators of guppies vary not only in their geographical distribution, but also in the fraction of their diets comprised of guppies, the frequency with which they take males and females, and in the size classes taken (Haskins et al. 1961; Seghers 1973; Endler 1978, 1983; Rodd and Reznick 1991; Pocklington and Dill 1995). All of these differences combine to produce a large amount of local variation in the ways in which predators affect the biology and evolution of guppies. The contribution of predators to geographic variation and behavioral plasticity is discussed in section 1.2 and chapters 4 and 5.

In addition to predators, the evolution and ecology of guppies may also be strongly influenced by another natural enemy, *Gyrodactylus turnbulli*, an external monogenean parasite (fig. 1.6). *Gyrodactylus* usually reproduces parthenogenetically and exhibits polyembryony, the development of embryos within embryos. As a result, parasite populations can grow very quickly on an individual guppy. *Gyrodactylus* have no intermediate hosts or resting stages in their life cycle and are unable to survive long off their host. Transmission occurs primarily through close proximity between an infected and an uninfected guppy. These features make this parasite ideal for experimental manipulations (e.g., McMinn 1990; Houde and Torio 1992). The guppy-*Gyrodactylus* system has been used for laboratory stud-

Table 1.1
Principal Predators of Guppies in Trinidad and Venezuela

Trinidad "Caribbean" Fauna	Trinidad "Mainland" Fauna	Venezuelan Fauna
Gobiidae	Cichlidae	Cichlidae
Eleotris pisonis	*Crenicichla alta*	*Crenicichla saxatilis*
Gobiomorus dormitor	*Aequidens pulcher*	*Crenicichla geayi*
Dormitator maculatus	*Cichlasoma bimaculatum*	*Caquetia kraussii*
		Aequidens pulcher
Cyprinodontidae	Lebiasinidae	*Astronotus ocellatus*
Rivulus hartii	*Hoplias malabaricus*	
	Hoplerythrinus unitaeniatus	Lebiasinidae
Mugilidae		*Hoplias malabaricus*
Agonostomus monticola	Characidae	*Hoplerythrinus unitaeniatus*
	Hemibrycon dentatum	*Lebiasina erythrinoides*
Centropomidae	*Astyanax bimaculatus*	
Centropomis unidecimalis		Characidae
	Cyprinodontidae	*Charax gibbosus*
Palaemonidae (prawn)	*Rivulus hartii*	*Acestrorhynchus microlepis*
Macrobrachium crenulatum		*Pygocenturus notatus*
		Cyprinodontidae
		Rivulus hartii

SOURCES: Endler 1978, 1983; Liley and Seghers 1975; Winemiller et al. 1990.

ies of host-parasite dynamics (e.g., Scott and Anderson 1984). There is some evidence for genetic variation in resistance to *Gyrodactylus* (Madhavi and Anderson 1985; Lyles 1990). Other parasites and pathogens of guppies may also be important in guppy populations but are not well studied (Lyles 1990). *Gyrodactylus* has been found commonly on guppies in several field surveys (Lyles 1990; G. Rosenqvist, pers. comm.), with prevalence up to 50% of individuals examined in some populations. Burdens are generally low (fewer than ten parasites per fish), possibly because heavy infections rapidly lead to death. About half of the guppies experimentally infected with *Gyrodactylus* in the laboratory are killed by the infection, while half eventually recover (pers. obs.). Infection with *Gyrodactylus* may play a role in mate choice by female guppies (see chapter 6).

1.2 Geographic Variation among Guppy Populations

The guppies of Trinidad and surrounding regions provide one of the best-known examples of geographic variation in any species. The physical distribution of guppies in many small streams results in numerous populations that evolve with various degrees of independence. Guppy populations in

Figure 1.6 Photograph of a guppy tail infested with *Gyrodactylus* parasites. Parasites are approximately 0.5 mm long. (Courtesy of A. M. Lyles.)

different streams are often strongly isolated from one another, depending on topography and the pattern of flow of streams into one another. Variation in predator assemblages and other environmental factors among guppy streams leads to parallel but independent evolutionary changes in guppy populations. The beauty of the guppy study system is that this geographic variation allows replicated observations of numerous populations and provides a natural laboratory for evolutionary studies. The potential for comparisons among populations has made the guppy system especially fruitful for sexual selection studies, allowing us to see how environmental variation affects the evolution of reproductive behavior and to look at the joint evolution of male secondary sexual traits and female mating preferences. Guppies are one of only a few species in which comparative approaches to sexual selection have been possible. Other notable examples of species in which geographic variation has been exploited in a similar way include threespine sticklebacks (Reimchen 1989; McKinnon 1995), house finches *Carpodacus mexicanus* (Hill 1994), and cricket frogs *Acris crepitans* (Ryan and Wilczynski 1988; Ryan et al. 1990a, 1992) (see later chapters for details). Most other comparative studies have examined patterns of sexual selection in groups of closely related species such as African killifishes *Diapteron* (Brosset and Lachaise 1995), swordtails and platyfishes *Xiphophorus* (Ryan and Wagner 1987; Basolo 1990a,b, 1995a,b), tropical frogs *Physalaemus* spp. (Ryan et al. 1990b; Ryan and Rand 1995), birds of paradise, Paradisaeidae (Pruett-Jones et al. 1990) and bowerbirds, Ptilonorynchidae (Borgia 1995). An advantage of working with intraspecific variation is that differences among populations are qualitative rather than quantitative and communication systems have not diverged enough to produce complete behavioral isolation.

Genetic Differentiation among Populations and Subpopulations

A few studies have examined the degree of overall genetic variation between and within guppy populations using biochemical (allozyme electrophoresis) and molecular (mtDNA sequencing and DNA fingerprinting) techniques (Carvalho et al. 1991, 1996; Shaw et al. 1991; Fajen and Breden 1992; Foo et al. 1995; Hornaday et al. 1995; Magurran et al. 1995; Sato et al. 1996). The initial results show significant divergence in sequence and allozyme frequencies among populations (fig. 1.7). In particular, there appears to be considerable differentiation between the two main south-flowing drainages in Trinidad's Northern Range, the Caroni and the Oropuche Rivers, between different north-flowing drainages, and between the south-flowing and north-flowing stream systems. These differences suggest that guppies may have become established in Trinidad through several independent colonization events (Carvalho et al. 1991; Shaw et al. 1991), but data from Venezuelan guppies are needed to verify this (Fajen and Breden 1992). The mtDNA differences between populations are consistent with divergence times of 100,000 or 200,000 years in some cases to as much as 600,000 years for the Oropuche drainage (Fajen and Breden 1992). These studies support the assumption that different populations are the result of largely independent evolutionary trajectories.

Within streams, the degree of isolation among different localities depends on the topography of the stream. Haskins et al. (1961) describe the population structure of guppies within streams as "beadlike," with relatively stable subpopulations at intervals along the length of streams and gene flow between subpopulations dependent on the presence or absence of barriers such as waterfalls. For example, the Aripo River of Trinidad, which flows from the Northern Range mountains south into the Caroni

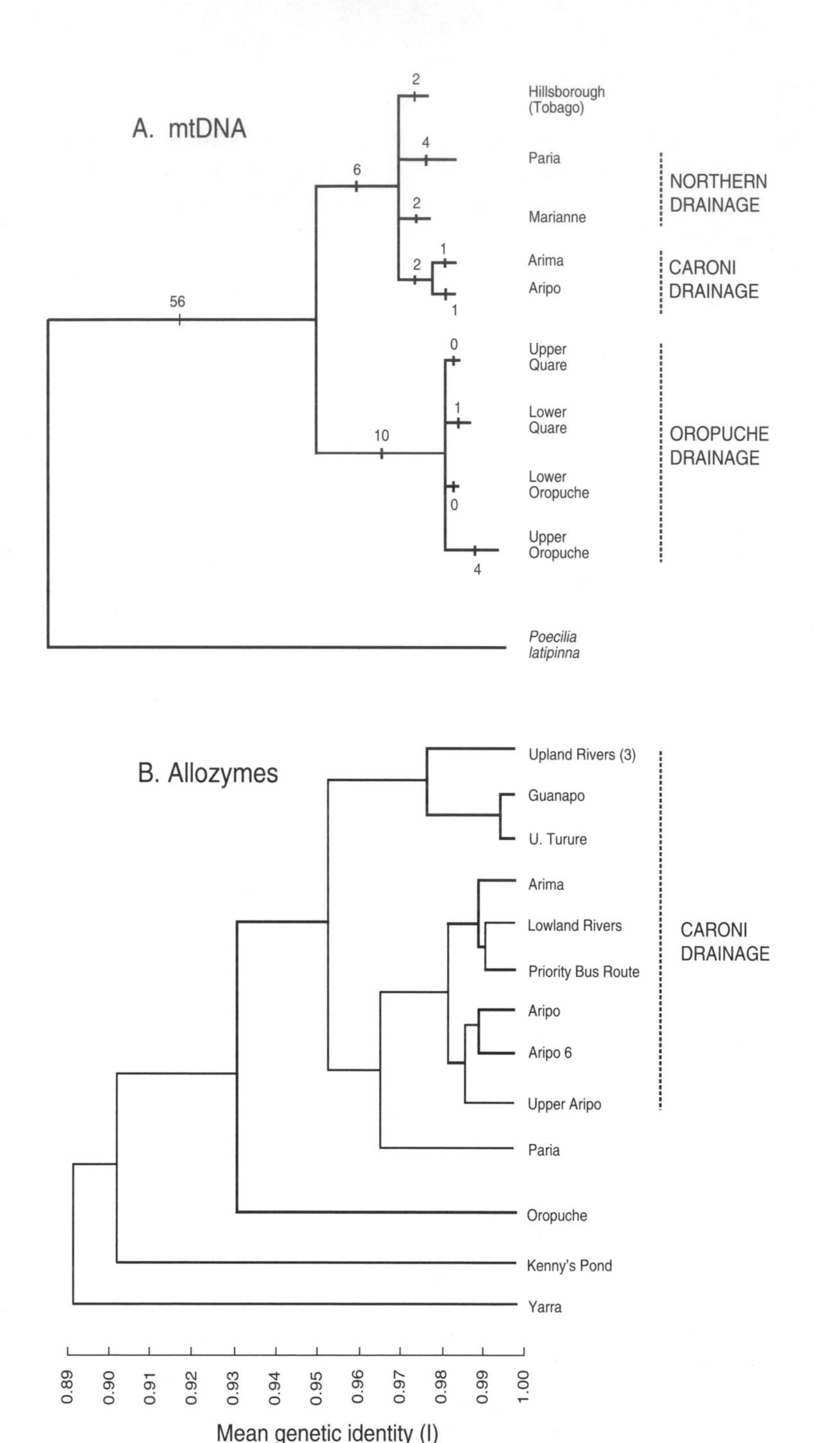

Figure 1.7 Phylogenetic relationships among guppy populations. (A) Relationships based on sequence comparisons of a 465 base-pair segment of the control region of the mitochondrial genome (adapted from Fajen and Breden 1992). Numbers indicate branch lengths from a strict consensus tree. (B) Relationships based on comparisons of frequencies at 23 allozyme loci (adapted from Magurran et al. 1995). Branch lengths do not correspond to those in (A).

River, has a major waterfall separating upstream from downstream populations of guppies. Guppies are unable to climb the waterfall, so there is little gene flow from below the waterfall to above the waterfall (Haskins et al. 1961; Carvalho et al. 1991). Guppies did colonize the river above the waterfall at some point, as they have in other rivers, though the mechanism for their transportation is not known. It has been suggested that guppies are occasionally transported by birds or wind storms, that they could have been present in Trinidad streams before the waterfalls were formed, and that historic changes in heights of waterfalls through geomorphological processes could allow colonization of headwaters (Endler, pers. comm.).

Even without abrupt barriers such as waterfalls, guppy populations appear to be fairly sedentary (reviewed in Magurran et al. 1995). Pools from which all the guppies have been removed sometimes remain empty for several weeks (Reznick, pers. comm.). During periods of heavy research activity in Trinidadian guppy streams, this void means that attempts to collect fish from favorite localities are sometimes fruitless. Researchers thus need to make an effort to protect these valuable guppy populations. (See Magurran et al. 1995 for further comments and recommendations about the conservation of guppies.) The existence of small differentiated subpopulations is supported by the observation of gene frequency differences between localities within streams with a pattern of variation consistent with "isolation by distance" models (Carvalho et al. 1991; Shaw et al. 1991, 1994; Magurran et al. 1995).

Haskins et al. (1961) attempted to document patterns of gene flow by releasing genetically marked guppies into streams. They introduced fish carrying a known marker (*maculatus*) into the native populations, then monitored the frequency of the marker for seven years after the introduction. In three such introductions, the marker was detected as far as 4.8 km, 8.8 km, and 9.8 km downstream of the introduction site within the monitoring period. In the last case, the *maculatus* marker traveled the full 9.8 km distance within the first 16 months. From these results, Endler (1977) estimated the gene flow distance (l) for guppies to be about 0.75 km per generation. A subsequent introduction of guppies into the Upper Turure River of Trinidad appears to have resulted in complete replacement of the original guppy populations (Magurran et al. 1992; Shaw et al. 1992; see discussion of the Turure introduction in the next section). Although guppies appear to be relatively sedentary in the short term, there is thus significant opportunity for gene flow along streams when barriers are not present.

The physical and ecological barriers to gene flow and the sedentary nature of guppies may be sufficient to account for the degree of evolutionary independence of different guppy populations. But a further intriguing possibility is that the sexual behavior of the guppies may also contribute to marked differentiation among populations. Mating preferences of females for local males (see chapter 6) may reduce gene flow and contribute to differentiation of populations. Also, to the extent that male mating success is strongly skewed by sexual selection processes, effective population sizes may be reduced, promoting rapid evolutionary divergence. This effect could be especially pronounced in low-predation populations where sexual selection may be strongest (Shaw et al. 1994). These ideas remain to be tested, however.

Geographic Variation and the Effects of Predation

Predation regimes vary within and between numerous localities in Trinidad and Venezuela, resulting in dramatic patterns of geographic variation in many characters (see Endler 1995 for a more complete compilation of effects of predation regime). This geographic variation is strongly influenced by the variation in predator assemblages (Strauss 1990; Endler 1995), but may also result from other environmental differences, from genetic drift, and also from independent processes of sexual selection. It is no simple matter to determine how various evolutionary processes interact to produce the geographic patterns we observe.

Not only do waterfalls such as that on the Aripo River act as barriers to the movements of guppies, but they also limit the distribution of the more dangerous predators (e.g., *Crenicichla alta*) to the downstream portions of rivers. Above the Aripo waterfall, guppies are subject to only minor predation, mainly on juveniles by *Rivulus* and *Aequidens*; but below the waterfall, adult guppies are at great risk of predation by *Crenicichla.* Differences in behavior, life history, and conspicuousness of color patterns are all associated with this sharp change in predation regime along the Aripo River (see next section). Introduction experiments involving transplants of both guppies and predators across the Aripo waterfall and elsewhere have confirmed that predation regime is indeed a major cause of differences in the guppies.

The color pattern of male guppies was the first character shown to vary geographically. The obvious explanation is that selective predation on conspicuous individuals could shape the evolution of color patterns and that different predation regimes would lead to differences in color patterns among localities. In laboratory tests, conspicuous males are at greater risk to predators than less conspicuous males (Haskins et al. 1961; Godin, pers. comm.), but surprisingly, females may be preferred by pike cichlids (Pocklington and Dill 1995). Early field studies by Haskins et al. (1961) showed that particular color pattern elements varied in frequency among localities. The variation documented by the Haskinses did not always correspond to differences in predation regime, but they did find that the genetic linkage of

Table 1.2

Linkage of Sb Color Pattern in Male Guppies from Above and Below the Aripo River Waterfall

Linkage of Sb Pattern	Above Waterfall	Below Waterfall
X	4	0
Y	8	33
X and Y	7	0

SOURCE: Data from Haskins et al. 1961

NOTES: Table gives numbers of wild-caught males with X, Y, and both X and Y linkage of the pattern. The population above the waterfall is subject to low levels of predation by *Rivulus*; the population below the waterfall is subject to higher levels of predation by *Crenicichla* and other species. Linkage of the *Sb* element was determined by breeding experiments in which X-linked color patterns were "developed" in female offspring with testosterone treatment.

color patterns differed between high predation and low predation. Particular color pattern elements appear to be inherited through both the X and the Y chromosome above the waterfall on the Aripo River but are inherited only through the Y chromosome below the waterfall (table 1.2). Haskins et al. (1961) explored the linkage of color patterns further using testosterone treatment of females from different populations to determine if unexpressed X-linked or autosomal patterns were present. The testosterone treatment "develops" any color pattern carried by the female. The results were consistent with the breeding experiments with Aripo guppies: fish from low-predation (upstream) localities had X-linked and autosomal color patterns, while fish from high-predation (downstream) localities had only Y-linked color patterns—high predation females never "developed" color patterns with testosterone. This difference in the genetic basis for color patterns between high- and low-predation localities is contrary to expectation (Endler, unpublished simulation results) and has never been adequately explained.

More extensive studies by Endler (1978, 1980, 1983) showed that quantitative characteristics of color patterns showed consistent differences associated with differences in predation regime even though allele frequencies at specific color pattern loci varied haphazardly from population to population (figs. 1.8, 1.9). The color patterns of males tend to be more conspicuous in populations with low levels of predation (localities with only *Rivulus* as predator), while color patterns tend to be less conspicuous where more dangerous predators (e.g., *Crenichicla*) are present. Similar variation in color patterns in Venezuelan populations has been documented by Winemiller et al. (1990). As in the Trinidad populations, guppies tended to be most conspicuous in streams with the least risk of predation.

The correlative results from sampling natural populations (Endler 1978, 1983) were strongly suggestive but needed experimental confirmation. In greenhouse experiments, Endler (1980) set up replicated artificial streams

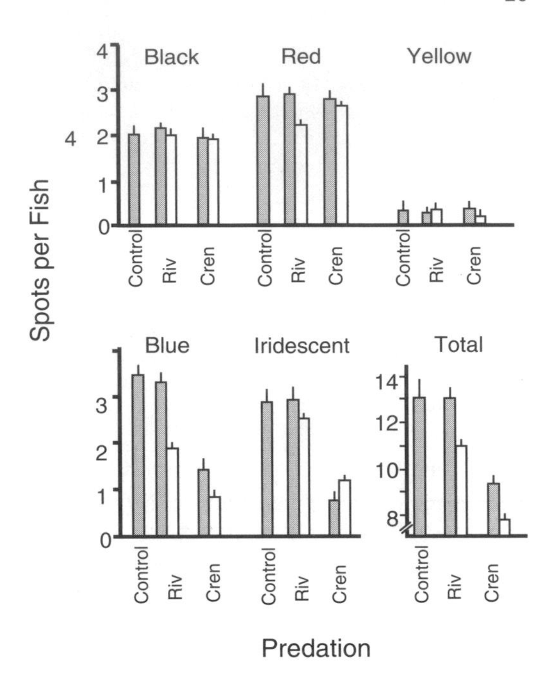

Figure 1.8 Effect of predation regime on numbers of color spots in guppy color patterns. Open bars: results from field surveys; closed bars: results from greenhouse experiments. Riv: populations with *Rivulus*. Cren: populations with *Crenicichla*. Control: greenhouse populations with no predators. Error bars indicate two standard errors. (Adapted from Endler 1980.)

containing *Crenicichla alta, Rivulus hartii*, or no predators (control) and two different gravel sizes, and he introduced a composite population of guppies derived from several Trinidad localities into each. As in the field populations, guppies in *Crenicichla* treatments evolved fewer and smaller spots than did guppies in *Rivulus* treatments (figs. 1.8, 1.9). Guppies in both predation treatments evolved less conspicuous color patterns with larger spot sizes in coarse gravel treatments and smaller spots in fine gravel treatments. The opposite occurred in control populations with no predators: guppies evolved greater conspicuousness with spot sizes that mismatched the gravel background, presumably as a result of sexual selection. A field experiment in which guppies from a high-predation population were introduced to a low-predation tributary lacking in guppies also confirmed the hypothesis of natural selection on color patterns by predators (Endler 1980). The fact that guppies seem to evolve to be as conspicuous as possible, given restrictions imposed by the local risk of predation, led directly to the argument that conspicuous color patterns must be favored in sexual selection and to the recent research program on sexual selection in guppies described in chapters 3–6.

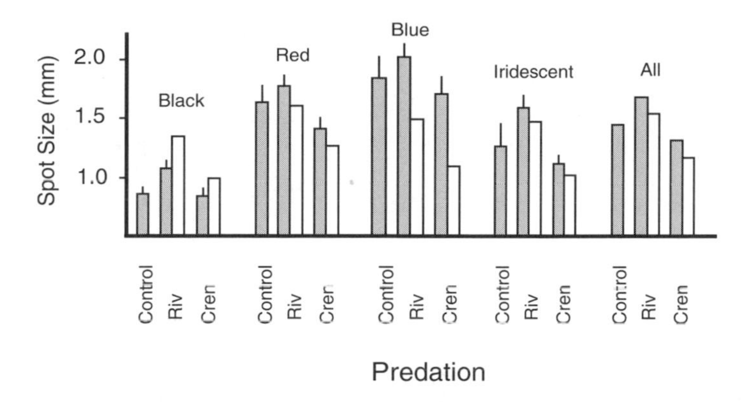

Figure 1.9 Effect of predation regime on sizes of spots. Labeling as in figure 1.8.

Life history traits also vary in parallel with differences in predation regime. Guppies from high-predation populations tend to produce many small offspring early in life and at frequent intervals , while guppies from low-predation populations produce larger but fewer offspring less often and begin reproducing later in life (Reznick 1982, 1989; Reznick and Endler 1982; Reznick and Bryga 1996; Reznick et al. 1996; fig. 1.10). These differences probably reflect the longer life expectancy of fish in low-predation populations. Large size at maturity may also be favored where *Rivulus* is the main predator and takes mainly small, immature size classes of guppies. Guppies in these populations may be selected to outgrow the predator before maturing (Seghers 1973; Liley and Seghers 1975; Reznick and Endler 1982). This may also contribute to the larger size of young guppies at birth in *Rivulus* localities (Reznick and Endler 1982). As with color pattern differences, two field introduction experiments (Reznick and Endler 1982; Reznick and Bryga 1987; Reznick et al. 1990) demonstrated that the variation in life history characters is indeed the result of natural selection by predators (fig. 1.10). These introduction experiments represent one of the strongest direct demonstrations of evolutionary change resulting from natural selection in natural populations.

The presence of large predatory fishes such as *Crenicichla* is associated with antipredator behavior of guppies. Behavioral changes in response to predators may be long-term, "hardwired" evolutionary changes, or they may be the result of plasticity in behavior, or both. When differences in response to predators persist in laboratory descendants of high- and low-predation guppy populations (e.g., Seghers 1973), we can conclude that this behavioral variation has a genetic basis.

Guppies from high-predation localities tend to form more cohesive schools, remain near the stream banks, and are more likely to inspect predators and respond to predators at a greater distance and with lower stimulus

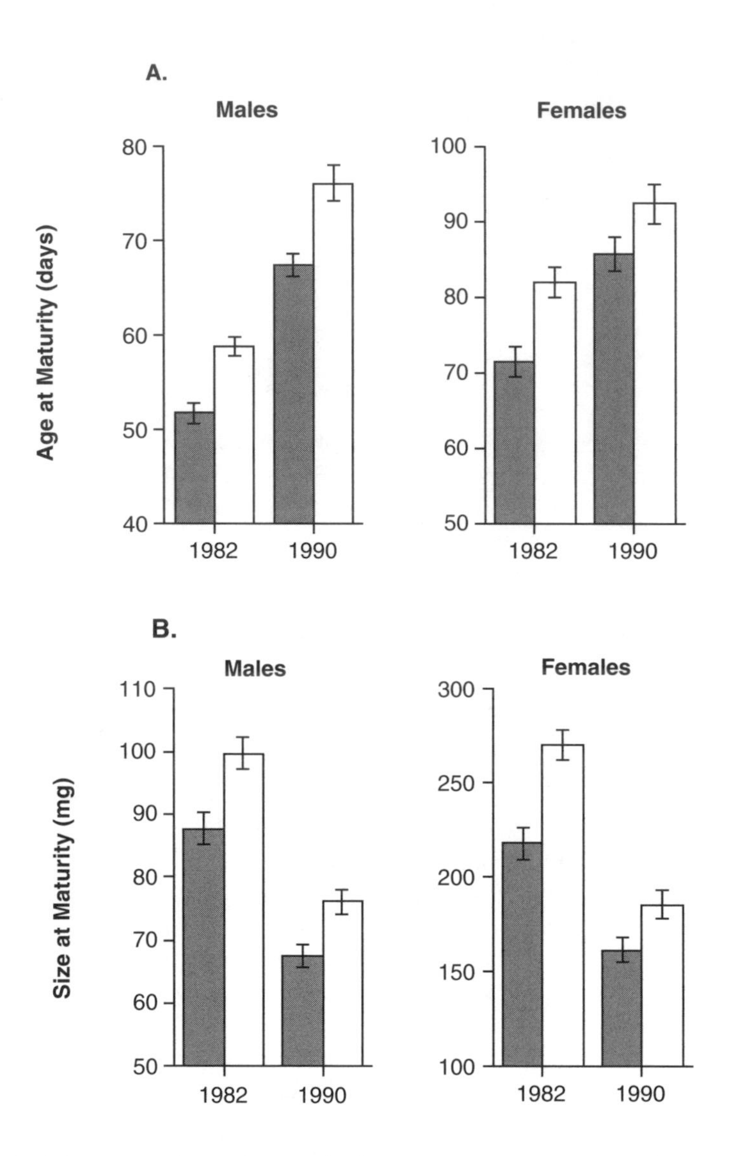

Figure 1.10 Differences in age (A) and size (B) at maturity in guppy populations in which the major predator is *Crenicichla* (shaded bars, high predation) or *Rivulus* (open bars, low predation). The 1982 data are from comparisons of naturally occurring fish from two localities on the Aripo River (Reznick and Endler 1982). The 1990 data are the results of an introduction experiment (Reznick et al. 1990); shaded bars are the original (high-predation control) population, and open bars are data for fish introduced from the original population to a low-predation *Rivulus* locality where guppies previously did not occur. Error bars indicate one standard error.

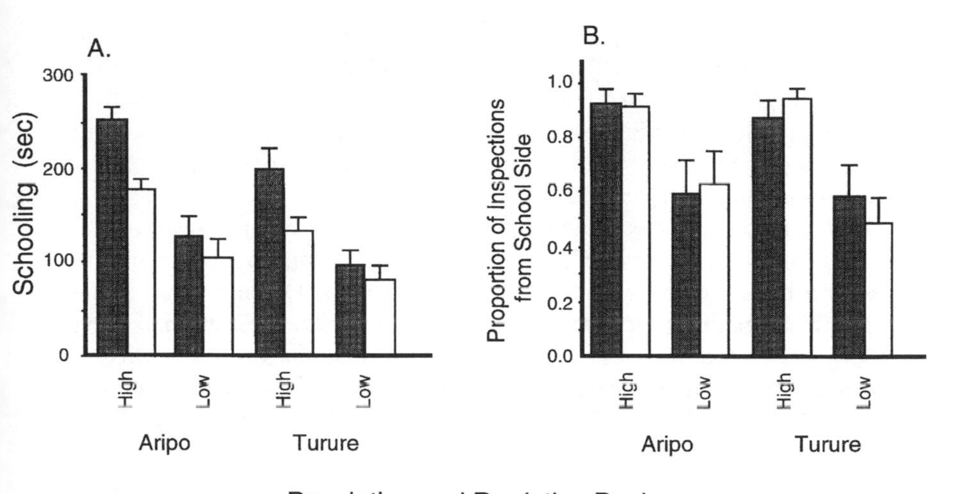

Figure 1.11 Effect of predation regime on antipredator behavior of male and female guppies in laboratory experiments. Comparisons are between fish from high- and low-predation localities in two Trinidadian rivers. Dark bars: females; open bars: males. (A) Fraction of time spent schooling. Test guppies were allowed to choose between a group of guppies on one side of the test aquarium (the school) and no guppies on the other side. (B) Proportion of predator inspections performed when test individuals were on the same side of the aquarium as other guppies, i.e., from the "school" side of the aquarium. Error bars indicate one standard error. (Adapted from Magurran et al. 1992.)

threshold than do guppies from low-predation localities (e.g., fig 1.11; Seghers 1973, 1974b; Breden et al. 1987; Licht 1989; Magurran and Seghers 1990a,b, 1991, 1994b; Magurran et al. 1992, 1993, 1995; Mattingly and Butler 1994; but see Seghers and Magurran 1995). For practical purposes in guppy research, the variation in response to predators means that it is possible to capture guppies from some low-predation sites with one's bare hands, while considerable patience and agility with two nets is necessary to capture guppies in most high-predation sites. Guppies from different populations also vary in the particular kind of behavior exhibited in response to piscine predators, ranging from weak avoidance to leaping out of the water (Seghers 1973). Although much of the variation in schooling behavior appears to be related to differences in predation intensity, there are also differences specific to particular drainages (Magurran et al. 1992, 1993, 1995; Seghers and Magurran 1995).

Populations also differ in response to aerial predators (Seghers 1974a). Seghers compared responses to a simulated aerial predator in descendants of fish from the Paria River (low level of aquatic predation) to that of fish from the Guayamare River (very high level of aquatic predation). The Paria guppies were more responsive, moving to deeper water and remaining still for longer than the Guayamare guppies. It may be that Paria guppies are subject to greater risk of predation from the air, e.g., by kingfishers (*Chloroceryle americanus*; see Endler 1978 for further discussion). A more likely interpretation is that Paria guppies are free to respond maximally to an aerial predator, while the response of Guayamare guppies is a compromise between the risk from the aerial predator and the risk of predation in the water column (Seghers 1974a).

Different populations of guppies respond to the actual presence of a predator in different ways. Upon detecting a predator, guppies exhibit inspection behavior (Magurran and Seghers 1990a; Dugatkin and Alfieri 1991a,b; Dugatkin 1992b; Dugatkin and Godin 1992a; Godin and Davis 1995), in which they observe and monitor the predator's behavior. Inspection appears to function in recognition of the predator and assessment of its potential threat (Magurran and Seghers 1990a; Dugatkin and Godin 1992a). Guppies from high-predation populations stay farther away from the predator when inspecting, and are more likely to stay near other guppies when inspecting than are low-predation guppies (fig. 1.11; Magurran et al. 1992). These differences persist in laboratory-reared fish and appear to be genetic. Wild-caught guppies seem to have specific familiarity with predators common in their own populations. They are most wary of their own predators and avoid the predator's mouth area to a greater degree than they do with unfamiliar predators (Magurran and Seghers 1990a). The degree to which these responses to particular predator species are genetically based is not known, however.

As with color patterns and life history traits, evolutionary changes in antipredator behavior as a result of differing predation regimes has been confirmed with an introduction experiment (Shaw et al. 1992; Magurran et al. 1992). In 1957, Caryl Haskins introduced 200 guppies from a high predation locality on the Arima River into a low predation locality in the upper portion of the Turure River, which was free of guppies at the time. Analysis of allozyme frequencies in 1991 revealed that the introduced population had colonized the upper Turure and had apparently migrated downstream and displaced the guppy populations in the high-predation parts of the stream. The reason for this displacement of original populations is not known. The result was a gene frequency profile for the whole Turure River similar to that of the original population from the Arima River, suggesting that the loci surveyed were selectively neutral (Shaw et al. 1992). However, predictable changes in behavior did occur that parallel the changes seen in the low-predation and high-predation parts of the Aripo River. By 1991 lower Turure (high-predation) guppies showed a greater tendency to school with other individuals and maintained a greater distance when inspecting a predator than did upper Turure (low-predation) guppies (fig. 1.11; Magurran et al. 1992). The introduced guppies appear to have undergone adaptive changes as they colonized parts of the river with different predation regimes.

These are undoubtedly only a small sample of the behavioral adaptations of guppies in response to the varying intensity of predation and species of predators. There are also clear effects of predation regime on feeding behavior (Fraser and Gilliam 1987; Godin and Smith 1988), and on courtship and mating behavior (chapters 4 and 5). Our understanding of how geographic variation affects any one aspect of guppies, for example sexual selection, is enhanced by the detailed information we have on other interrelated aspects of their biology.

1.3 Issues in Sexual Selection and What Guppies Can Tell Us

Interest in the biology of guppies has focused on their conspicuous color patterns and on their sexual behavior ever since the species was first known. The presence of a conspicuous color pattern was more than likely the initial motivation for studies of sexual selection in guppies and has led to the accumulation of a large body of literature in recent years. We now have information on topics ranging from patterns of evolution across guppy populations to patterns of sensitivity of retinal cells of individual guppies. In a similar way, Darwin (1871) was motivated to develop the theory of sexual selection by the existence of elaborate sexual ornaments in a variety of species. The resulting explosion of theoretical and empirical research on sexual selection in the past quarter century is recognized as a major thrust of modern evolutionary biology.

The remainder of this chapter summarizes the main issues in sexual selection that have been, or could be, approached with guppies. The details of the studies themselves are given in chapters 3 through 6. A number of excellent reviews and syntheses of sexual selection theory have appeared recently (e.g., Pomiankowski 1988; Grafen 1990a,b; Ryan 1990; Michod and Hasson 1990; Balmford and Read 1991; Iwasa et al. 1991; Kirkpatrick and Ryan 1991; Reynolds and Gross 1990; Pomiankowski et al. 1991; Maynard Smith 1991; Price et al. 1993; Andersson 1994; Iwasa and Pomiankowski 1994; Bakker and Pomiankowski 1995; Johnstone 1995; Møller 1994b; Iwasa and Pomiankowski 1995), so I will not attempt a definitive review of sexual selection here.

Sexual Selection and the Evolution of Sexual Ornaments

The color pattern of guppies is a perfect example of the kind of trait that presented a difficulty for Darwin's (1859) theory of natural selection, and that could be explained by his (1871) theory of sexual selection. A conspicuous color pattern increases a guppy's risk of being preyed upon (see above), so the obvious explanation for the existence of this trait in populations is that it provides an advantage in sexual competition. The initial goal of the first studies of sexual selection in guppies was to test this hypothesis and to determine which mechanism of sexual selection—male-male competition or female choice—was operating. Guppies have provided an especially good opportunity to investigate female choice. Chapter 3 describes studies testing the basic hypothesis that sexual selection favors conspicuous colors and other traits in guppies and examines the effects of female choice and male-male competition on male mating success and their contributions to sexual selection.

It now appears that the nature of the balance between sexual selection favoring conspicuousness and natural selection by predators favoring inconspicuousness is more complex than was originally believed (see fig. 7.1). The display behavior of male guppies also shows a great deal of plasticity, reflecting opposing costs and benefits of courtship under varying conditions (chapter 4). Rather than representing a unidimensional equilibrium between two opposing agents of selection, color patterns now appear to reflect a more complex interplay between the spectral properties of the colors themselves, condition-dependent expression of the colors, effects of the environment on how light reflected by color patches is transmitted to the female, and the sensory capabilities of guppies and their predators (see chapters 5 and 6). Finally, the evolution of mating preferences of females and of preferred traits of males interact in complex ways (chapter 6). (See fig. 7.1 and chapter 7 for an overview of the evolutionary and ecological relationships now known to affect sexual selection of guppies.)

Male Courtship Behavior

The obvious preoccupation of male guppies with the pursuit and courtship of females is as striking as the conspicuous color patterns. The mate searching strategies and courtship behavior of male guppies have been extensively investigated (see chapter 4). Studies generally view mate searching and courtship as a life history problem and seek to understand how variation in male behavior is related to variation in factors such as risk of predation, food availability, attributes of females, and the social environment. Variation in the courtship behavior of males may be expressed as genetically hardwired differences among populations, or as conditional plasticity in the behavior of individuals.

How Do Females Choose?

Recently, studies have begun to consider the behavioral and sensory details of female mating preferences and the possibility that preferences are shaped by costs associated with mate choice (see chapter 5). Despite frequently made assumptions about the nature of female choice in theoretical models and even explicit modeling of the fitness consequences of different choice rules, there is surprisingly little information about exactly how females express their mating preferences (but see Eberhard 1996). Studies of guppies have provided some information about how females respond to male phenotypes and rules they use in mate choice, but this is an area where much more work is needed.

Guppies have been especially useful for demonstrating variability in mating preferences both between and within populations (see also chapter 6). Population differences in female choice demonstrate that the mating preferences of guppies are genetically variable and are subject to evolutionary change. Recent studies have examined phenotypic plasticity in preferences in response to variables such as predation regime and social environment.

At a more proximate level, mating preferences are being investigated in terms of underlying sensory processes. Information about the visual system of female guppies is providing clues about how mating preferences come about, and studies of the lateral line system are investigating the possibility of vibration as a mode of sexual communication.

Evolution of Female Choice

Given that female guppies seem to have mating preferences based on the color patterns of males and that preferences show genetic variation and phenotypic plasticity, the next step is to ask how and why these preferences might evolve (see chapter 6). Mating preferences are the cause of sexual selection on male traits such as color patterns, so the evolutionary dynamics of preferences and preferred male traits are intimately interrelated. Much theoretical work has been devoted to understanding evolution of mating preferences, but empirical results have been slow in coming. Work with guppies has provided information about the applicability of various theoretical models for the evolution of female choice. Several studies support the idea that female preferences are based on condition-dependent traits that might indicate heritable "good genes." Other studies have shown that male color patterns and female preferences appear to evolve in parallel in different guppy populations, possibly as a result of genetic correlation between these traits within populations.

1.4 Summary

Guppies have provided a unique opportunity for studies of sexual selection because of their color patterns, the accessibility of their native streams in Trinidad, the ease of laboratory breeding and maintenance, and the valuable groundwork on their evolution and ecology resulting from the pioneering work of the Haskinses, of Endler, and others. The color patterns of guppies, given, as we shall see (chapters 2 and 3), their behavior and mating system, are a perfect example of the kind of secondary sexual characters that can be explained by Darwin's (1871) theory of sexual selection. Males have conspicuous color patterns and sexual displays despite clearly demonstrated costs imposed by risk of predation. Furthermore, we have the added advantage that guppy color patterns are genetically based and are highly polymorphic, allowing easy recognition of individuals for behavior studies. The high heritability of color pattern differences even permits us to assign paternity in mating success studies (see chapter 3 and Appendix). Finally, the division of guppies into more or less isolated populations in different streams means that studies of sexual selection can be extended from examination of processes within a single population to comparative, replicated studies of evolutionary dynamics occurring among populations. Variation in physical and biotic factors among streams contributes to the variation among these independently evolving populations of guppies. These factors interact with the sexual selection processes going on in each stream in ways that give us insights into the evolutionary dynamics of mating preferences and color patterns (see chapter 6). Guppies are unique among the species in which sexual selection has been studied intensively in that we have been able to make specific predictions about evolutionary outcomes based on different populations. There are thus many advantages, and relatively few limitations of the guppy study system, for investigations of sexual selection.

2 Reproductive Biology and Sexual Behavior

This chapter outlines the reproductive biology and sexual behavior of guppies. I have not attempted to review all of the early, sometimes conflicting, studies of sexual behavior in guppies. Rather, my intention is to give a consensus view of how reproduction in guppies works as background for the studies of sexual selection and mate choice detailed in later chapters. Where appropriate, I have included information from my own unpublished observations. I have also tried to give some practical information that may be of use in designing future experiments (see the Appendix for further details on experimental methods).

Guppies, like other poeciliid fishes, are livebearers with internal fertilization. Females carry the developing embryos until the yolk sacs are absorbed, and then give birth to live young. Females can store sperm from one insemination to fertilize several successive broods of young. These characteristics of poeciliid reproductive biology mean that a single pregnant female, carrying young fathered by several males, can found a new population or recolonize an area after a catastrophic event such as a flood. Generation times of guppies are short: as little as two months between fertilization of a zygote and first mating. Female guppies can produce large numbers of young—twenty or more young per litter for a large female. The short generation time combined with large litter sizes gives guppies the potential for rapid population growth, as those who have kept guppies in home aquaria are well aware. The guppy has been aptly termed the "Millions Fish" by Trinidadians.

2.1 Mechanics of Reproduction

In guppies, ova are matured in nonoverlapping batches. A new batch of ova matures shortly before the birth of each successive litter of young and the eggs are fertilized immediately after the young are born. Consequently, female guppies are sexually receptive, show behavioral responses to males, and are most likely to mate within a few days of giving birth to a litter of young, but are unresponsive to females at other times (fig. 2.1).

Liley (1966) conducted experiments in which he observed the behavior of females in response to males over time, starting from the day males were first introduced to virgin females (fig. 2.1A) or from the birth of a litter of

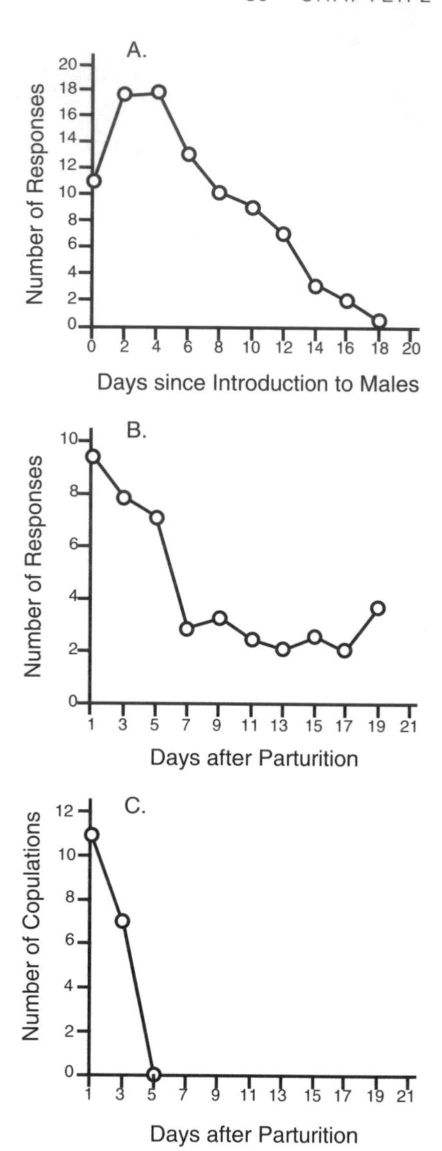

Figure 2.1 Decline over time in frequency of sexual behavior (glide responses or copulations) in virgin females from the day of first introduction to males (A) and in non-virgin females following the birth of a litter of young (B and C). (Based on data in Liley 1966.) See text for methods and description of females' glide response.

young (fig. 2.1B,C). Females were exposed to males for a period of fifteen minutes every other day. Virgin females and recent postpartum females copulated and showed high rates of sexual response to males initially (see below for description of females' glide response), but the frequency of these behaviors declined over the course of several days.

In my own experiments, virgin females that are introduced to males and are subsequently housed with them continuously usually copulate almost immediately on introduction and remain responsive to males for two or three days before becoming unresponsive. Mature virgin females that have never been exposed to males are probably unnaturally responsive to males. This may be convenient for observing their sexual behavior over several days in experiments. On the other hand, the fact that naive virgins mate so quickly when they first encounter a male suggests that they may be indiscriminate in their choice of mates relative to females that have matured in the presence of males.

Receptive females appear to produce a pheromone that is detected by males (Crow and Liley 1979; Meyer and Liley 1982; Reznick, pers. comm.). Males increase their courtship rate in response to water used to house a receptive female even when the female is not actually present. Males could use their ability to detect the female pheromone both to locate females, especially in very low density populations, and to discriminate receptive from unreceptive females. The pheromone produced by receptive females may also lead to increased competition among males.

Guppies are considered lecithotrophic (Constantz 1989; Reznick and Yang 1993), meaning that the young are nourished before birth primarily by the yolk. Some other poeciliid species are considered matrotrophic, meaning that females supply additional nutrition directly to the embryos during development. Most lecithotrophic poeciliids, like guppies, mature only one brood of young at a time, while most matrotrophic species have superfetation: they carry overlapping broods of embryos at more than one stage of development at the same time (Constanz 1989; Reznick 1989), and are consequently more continuously fertile than guppies. Developing guppy embryos are carried within the female for three to four weeks and then are born when the yolk sac is mostly absorbed.

Females guppies can store sperm and produce several successive litters of young even in the absence of males (Winge 1937). New sperm from recent matings are more likely to fertilize ova than stored sperm, and sperm from multiple matings during the female's fertile period are likely to produce multiple paternity (Winge 1937; Rosenthal 1951; Hildeman and Wagner 1954). Almost all females taken from the field or from group aquaria produce multiply sired litters of young, judging from the color patterns of the sons, which closely resemble those of their fathers (see chapter 2). Most litters produced by females with access to several males appear to be sired by two males, a few by three or more males (Houde, unpublished data; Luyten and Liley 1991).

A male transfers sperm to the female using his gonopodium, the anal fin modified as an intromittent organ. Three rays of the anal fin are thickened and elongated to form the rodlike gonopodium (fig. 2.2). The fin rays form a tube or channel allowing passage of sperm bundles (spermatozeugmata), and the gonopodium is tipped with sensory and gripping structures (Clark

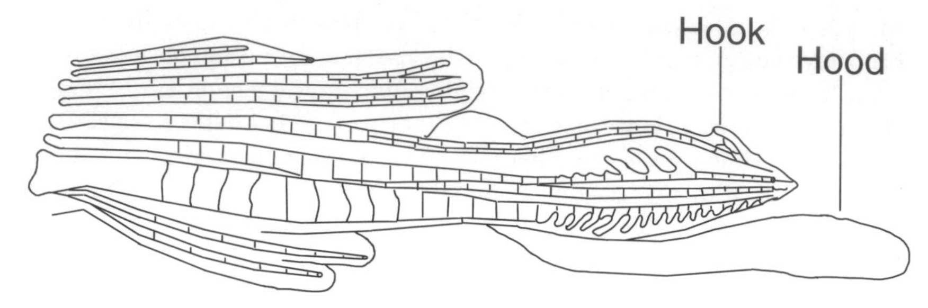

Figure 2.2 Diagram of the gonopodium of the male guppy. (After Clark and Aronson 1951.)

and Aronson 1951; Constantz 1989). Male courtship in combination with female responsive behavior (see details below) positions the male and the female so that the male can swing his gonopodium forward, insert it into the female's genital opening (a "gonopodial thrust"), and transfer sperm (fig. 2.3). A male can position himself near the female and insert his gonopodium without her cooperation, resulting in a "sneak" copulation (see details in sec. 2.3).

2.2 Ontogeny

Under laboratory conditions, male and female guppies begin to differentiate visibly at about four weeks of age. At this age, females can be distinguished from males by the presence of speckling or dark coloration in the anal region (fig. 2.4). The pigmentation is actually on the wall of the abdominal cavity rather than on the outer surface of the fish. This distinguishing feature is best seen in anaesthetized fish viewed under magnification. Sexes can be separated using this character well before males mature and become capable of inseminating females; the gonopodium and color pattern of males do not begin to develop until somewhat later. The anal fin of males typically begins to differentiate at five to six weeks of age, and the color pattern begins to appear by the time the gonopodium has become distinctly rod shaped.

In raising guppies for mate choice experiments it is often useful to be able to judge if a male has matured enough to have inseminated the females he is housed with. This can be determined by examining the "hood," one of the sensory structures at the tip of the gonopodium (fig. 2.5). The hood is

---

Figure 2.3 (*opposite page*) Sexual behavior of guppies. *Top:* Male following female. *Middle:* Male in position for display to female. *Bottom:* Male displaying to female (sigmoid display). (Photos by Peter Nicinski and Jonathan Weiland.)

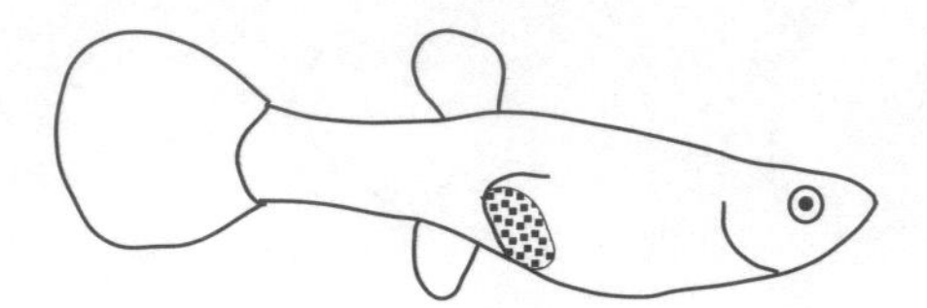

Figure 2.4 Diagram showing black speckling in anal region of females. This pigmentation is diagnostic of females even before they can be distinguished from males on the basis of gonopodium development.

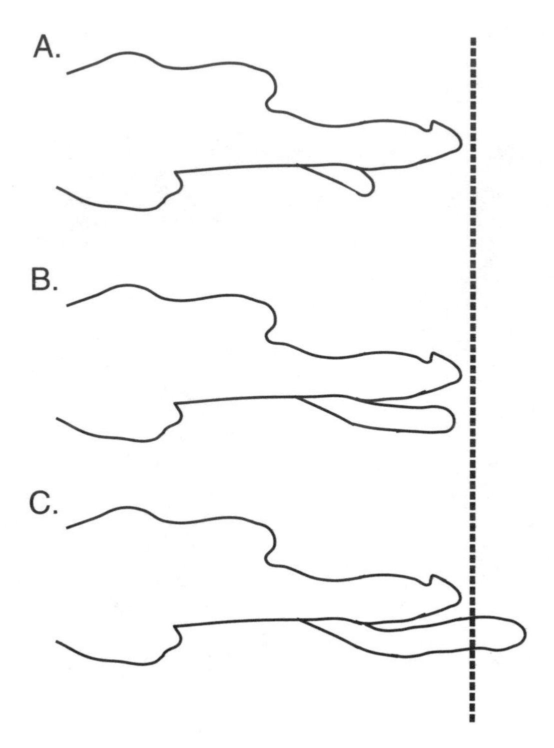

Figure 2.5 Stages of development of the gonopodial hood. (A) and (B) Males not capable of inseminating females. (C) Male capable of inseminating females. See text for explanation.

a protuberance on the ventral side of the gonopodium that elongates during sexual maturation from a small initial bulge to extend beyond the tip of the gonopodium proper. It appears to have a sensory function but is not essential for successful insemination (Clark and Aronson 1951). Experience in rearing guppies in my laboratory has shown that males are capable of inseminating females once the hood has grown even with or beyond the tip of the gonopodium (fig. 2.5). Females housed with males at earlier stages of development, in which the hood does not yet extend beyond the tip of the gonopodium, do not become pregnant and can be considered virgins even if the males have begun to show their color patterns.

Males begin to show sexual behavior as the gonopodium differentiates. They follow females and thrust at them with the gonopodium even before the color pattern has begun to develop. Males begin to perform courtship displays as the color patterns are elaborated. There seems to be no detailed description of the early ontogeny of the sexual behavior of males in the literature (but see Rodd 1994 for postmaturation ontogeny of male sexual behavior). It would be particularly interesting to compare the ontogeny of courtship behavior among populations in relationship to the ontogeny of color patterns. Differences in risk of predation would be expected to affect both sexual display and color patterns, but how does variation in the risk of predation affect the relative timing of development of these characters?

Female guppies reach sexual maturity at about the same age as males. Females may actually mate before their first ova mature: young virgin females in the laboratory often seem to take several weeks longer than the usual interbrood interval of three to four weeks to produce young after their first mating. This suggests that these females may be mating before they have any ripe ova and then store the sperm until the ova mature.

2.3 Sexual Behavior

The sexual behavior of male and female guppies has been described most thoroughly by Baerends et al. (1955) and Liley (1966). The reader is referred to these studies for full ethological details. Earlier authors (reviewed in Clark and Aronson 1951; Baerends et al. 1955; Liley 1966) reported conflicting information about the sexual behavior of guppies. This may be because of the use of domesticated strains of guppies that may have lost some elements of sexual behavior, and because obvious female sexual responses are rarely observed in established aquarium groups, although they are easily observed with virgin females.

Male Courtship

A male guppy can inseminate a female with her cooperation, in a "true" copulation which is preceded by courtship, or without courtship or her cooperation using a gonopodial thrust to achieve a sneak copulation (Clark and Aronson 1951; Baerends et al. 1955; Liley 1966; but see reviews of conflicting earlier literature in these references). The courtship display of

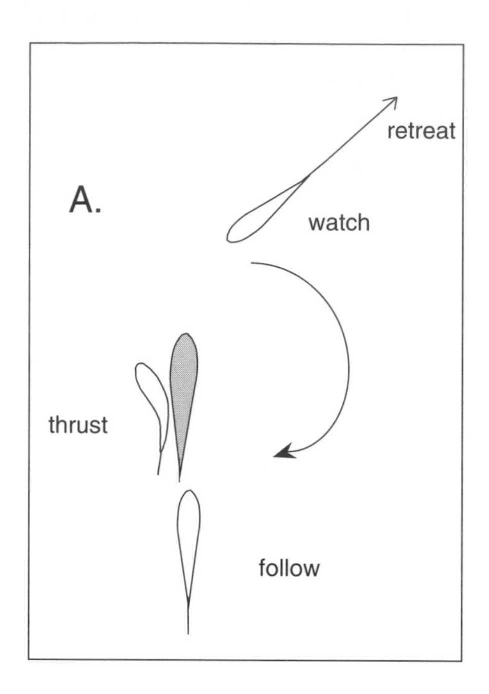

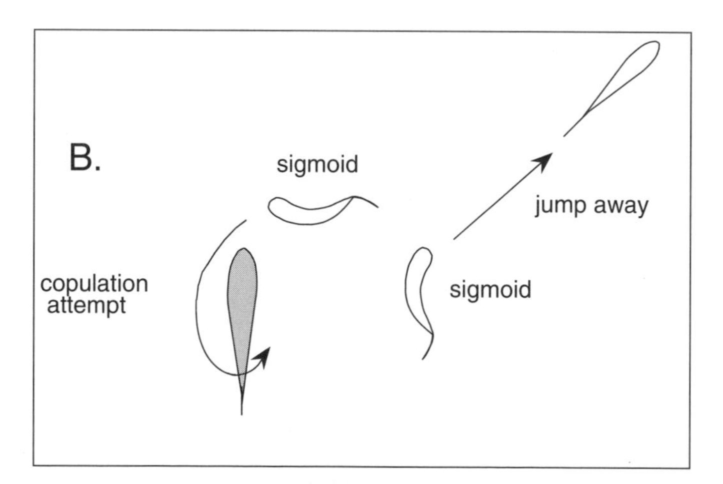

Figure 2.6 Diagram showing alternative mating tactics of guppies. (A) Thrust. The male (white) watches from a position in front of the female (shaded); from here he may retreat or move behind the female and follow her. From his position behind the female he may approach from below and to one side and attempt a sneak copulation by performing a gonopodial thrust. (B) Courtship display. The male performs a sigmoid display to the side or in front of the female; he may circle around her and attempt to copulate or he may jump away from her. (Modified from Liley 1966.) See text for additional description of behaviors.

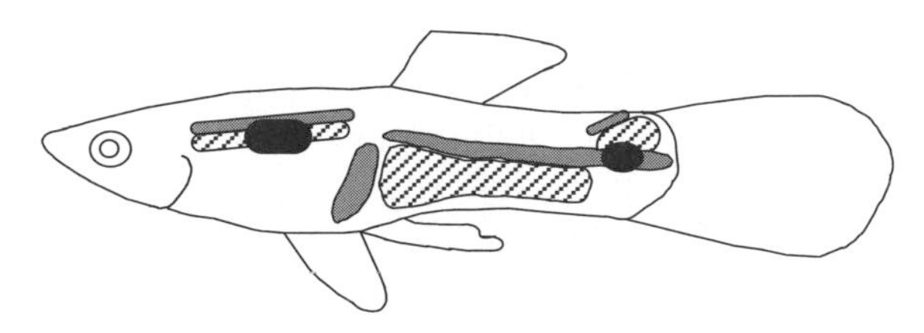

Figure 2.7 Elements of guppy color patterns expressed during courtship (example is one of many possible color patterns). Diagonal barring: orange spots; black: permanently black spots; gray shading: black spots expressed facultatively during courtship.

male guppies is termed the "sigmoid" display (Baerends et al. 1955; Liley 1966), named after the S-shape of the male's body as the display is performed (fig. 2.6). Males spend much of their time waiting in front or slightly to one side of the female, seemingly for an opportune moment to display. They generally display only when the female has stopped moving or has slowed sufficiently for him to move in front of her. A displaying male bends his body laterally into a sigmoid posture and quivers stiffly. Sometimes the quivering is brief and subtle. When the display continues for a longer time, the male may begin to pulse his body more slowly, moving several millimeters up and down in the vertical axis. The caudal and dorsal fins often remain closed during short displays, but may be spread in longer displays. Displays with open fins appear to be of higher intensity than displays with closed fins (Baerends et al. 1955; Liley 1966; Farr 1980a).

Courtship is usually accompanied by changes in the male's color pattern (fig. 2.7). Baerends et al. (1955) describe changes in color patterns in response to changes in the motivational state of the male and the size or receptivity of the female. Their study can be misleading, however, because their observations are based on only one or a very few different color patterns that undergo very specific changes with intensity of sexual behavior. In collections of guppies from wild populations, almost every color pattern is different and undergoes slightly different changes during courtship. In general, the black spots in male color patterns grow larger during courtship, and horizontal black lines appear on the body along the contours of the colored spots, but colored spots themselves do not change during courtship. Colors other than black show little if any change during courtship. The black lines seem to accentuate the spots, especially those with orange-red carotenoid colors. Some guppies I have observed have the ability to enlarge only a subset of potential black areas at one time, so that, for example, an anterior spot may dominate the color pattern at one time while at another a posterior spot may be the eye-catching feature of the color pattern. These males seem to be able literally to change their spots and dramatically alter their appearance to females. It is tempting to make the inference that whatever has led to the extreme polymorphism in color patterns between individuals has also favored this sort of polymorphism within individuals. Studies of the ability of individual males to control the expression of their color pattern and effects on female choice would be worthwhile.

This initial phase of the male's display may proceed to copulation, or the male may suddenly jump several centimeters away from the female (fig. 2.6). The function of this jump appears to be to try to lead the female away from her current position (Baerends et al. 1955) or to test her responsiveness. Display sequences are often interrupted by other males, so it is to a male's advantage to lead a female away from a crowded area. Males also tend to avoid displaying when other fish are nearby. Nevertheless, copulations can actually be "stolen" if an intruding male can move in at the crucial moment of the courtship sequence and thrust at the female.

Female Sexual Response

The courtship sequence does not proceed beyond the male's initial displays if the female is unresponsive. Unresponsive females appear simply to ignore the male's displays and continue feeding or searching for food. A responsive female orients toward the male so that she appears to be looking at him, and then glides smoothly toward him as he displays.

Many earlier studies failed to note any positive response on the part of females, perhaps because of the particular domesticated and laboratory strains they used. These studies generally did not use virgins, or used inbred and domesticated strains of guppies that may have lost elements of the full behavioral repertoire. Clark and Aronson (1951) characterized the sexual response of females only negatively, as a failure to flee from males, while Baerends et al. (1955) noted occasional positive responses in which females approached males. In early reports, every gonopodial contact was assumed to be a copulation, and copulations involving cooperation by the female were not recognized. Liley (1966) was the first to describe the responsive behavior of female guppies in detail (see also Kadow 1954).

The gliding motion of the female's sexual response is characteristic and is easy to distinguish from normal jerky swimming movements (Liley 1966). Gliding females seem to move with a rigid posture, moving only their fins. Sometimes the female's body is arched laterally toward the male as she glides. I refer to this gliding approach to a displaying male as a "sexual response." Liley (1966) suggests that this gliding movement may minimize the male's tendency to be frightened by the approach of the female, who is usually larger than the male. Females do not normally approach males in other contexts except in aggressive encounters in which the female is almost always the aggressor. The gliding response of females is thus a clear indication of the female's sexual interest in the male and has been used to assess female mating preferences (see, e.g., Houde 1987; Reynolds and Gross 1992; see Appendix for details on methods). The responsiveness of the female may affect how often the male displays and whether he performs only a brief, low-intensity display with closed fins or proceeds to perform a more high-intensity, open-fin display (Baerends et al. 1955). Males most often perform low-intensity, closed-fin displays when they first encounter a female and are testing for responsiveness. Once a female responds and begins to glide toward the male, he is more likely to intensify the display and spread his fins.

Even when the female responds to the male, the sequence of behavior does not necessarily proceed to copulation. The female may swim away from the males at any point, and the male may terminate the display himself or stop courting the female entirely. Most initially responsive females eventually lose interest in a given male before courtship advances to the point of copulation. When the female stops responding to a male's displays, the male may attempt a sneak copulation, then eventually gives up and moves on to another female.

COPULATION

If the courtship sequence does continue beyond the female's initial response, the male begins to move, still displaying, in a circle around the female. The female turns as the male displays around her. The male then swings his gonopodium forward and attempts to insert it from below and behind the female (fig. 2.3). If the male is successful in making contact, the pair rotates around each other rapidly. The male may be able to complete copulation immediately, whereupon he breaks away from the female almost violently. In other instances the pair may circle each other several times before insemination occurs, or they may separate with no successful insemination.

After a successful copulation, the male jerks his whole body up and forward, several times rapidly and then with decreasing frequency for up to several minutes. A jerking male seems to have a violent case of the hiccups and usually follows and appears to guard the female closely for several minutes. He may jerk in front of the female but does not perform any sigmoid displays. Males resume displaying after a refractory period that lasts as long as one hour.

The significance or function of the jerking behavior of male guppies is not clear, but it provides a useful indication of whether or not an observed gonopodial contact between a male and a female has resulted in successful insemination. Based on his own observations of virgin females and a review of earlier literature, Liley (1966) concluded that contact not followed by jerking never leads to insemination and pregnancy of the female, while almost all copulations followed by jerking lead to successful insemination and birth of young. Many possible functions of jerking have been suggested. One possibility is that jerking functions to resupply the gonopodium with sperm (Bowden 1969, cited in Constantz 1989). Jerking could also be a signal to other males or to females. Another idea is that jerking could dislodge any external parasites transferred to the male during sexual contact. If this were the case, however, males would also be expected to jerk in cases where prolonged contact occurs but does not result in successful insemination.

Sneak Copulations

The sequence of male courtship and female response that leads to copulation probably has evolved as a mechanism for facilitating encounter, contact, and insemination between a receptive female and a male who could be frightened by the close approach of a female. On the other hand, sneak copulation attempts occur fairly frequently as males follow and court females, especially when females are not responsive to a male's displays. Most sneak copulation attempts do not result in transfer of sperm. In early reports, all gonopodial contacts were assumed to result in insemination, and copulations involving courtship and cooperation by females were not recognized. Experiments by Clark and Aronson (1951) demonstrated that inseminations are rare relative to the rate of gonopodial thrusting (see also Luyten and Liley 1991; Kodric-Brown 1992). Most successful inseminations are preceded by courtship displays. Successful sneak copulation attempts can be identified by jerking behavior of the male (Liley 1966). Most successful sneak copulations that I have observed have occurred when the female was frightened, responding to the display of another male, or otherwise distracted. In most cases females actively avoid sneak copulations by darting away as soon as they detect the approach of the male and resuming their previous activity some distance away (Magurran and Nowak 1991).

2.4 Mating System

The mating system of guppies is best termed promiscuous. Males devote a large fraction of their daily time budgets to courtship and searching for females, and are especially active around dawn and dusk (Endler 1987). Males can potentially mate several times each day if receptive females are available. Females appear to mate with two or three different males each time they are receptive (see evidence on multiple paternity above). Most sexual behavior observed in natural social groupings consists of mate searching and courtship displays of males and avoidance of males by females. The mating system appears to be mainly one of female choice in that females are usually in control of when and with which male they mate. There is relatively little overt sexual competition among males (see section 2.5 and chapter 5).

In natural social groupings, males approach female after female, courting each one in succession. Since female guppies are only fertile when they first reach sexual maturity and for only two or three days out of each 25–30-day reproductive cycle, only about 10% of the females a male encounters are likely to be receptive. Females at other stages of their reproductive cycles generally ignore the courtship of males. The courtship activity of a male guppy is therefore an exercise in searching for receptive females that resembles a problem in optimal foraging. Males must locate and approach females rapidly, determine if each is receptive, and then move on to find other females. Courting males generally move quickly among unreceptive females, performing only one or a few displays to each before moving on. The courtship displays a male performs when he first encounters a female function mainly to test her responsiveness. Males become more persistent in their courtship when a female responds to a display, and may follow a responsive female for several to many minutes.

A courting male spends most of his time following females as they search for food. When the opportunity arises, the male may either perform a display or, more rarely, attempt a sneak copulation. Farr (1975) reported that males display up to twenty-seven times per ten minutes in Trinidad streams, and perform up to five gonopodial thrusts per ten minutes. These rates of display and thrusting are similar to those in my laboratory observations with stocks of guppies recently brought into the lab from Trinidad, but Farr (1980b) observed somewhat lower display rates in observations with old laboratory strains. Chapter 4 describes patterns of variation in rates of courtship display.

2.5 Aggression and Dominance

Aggression plays a very limited role in the social system of guppies. Although aggression and agonistic displays are in the repertoire of guppy behavior (see Baerends et al. 1955; Liley 1966), they occur only rarely in a sexual context. Relatively few poeciliids appear to have mating systems based upon aggressive interactions and dominance relationships among males (Farr 1989). These seem limited to species in which populations are at low enough density that stable relationships between individuals can be maintained (Farr 1984). By contrast, intermale aggression is an important component of the mating system of sticklebacks (Bakker 1994; Rowland 1994).

In most cases, males seeking females to court are simply reluctant to approach a female that is already being courted by another male. Males move from female to female sufficiently often that there are usually unaccompanied females readily available. Occasionally, an intruding male will persist in attempting to court a female that is already attended by another male. The original male then moves closer to the female and appears to "fend off" the intruding male, sometimes by lashing his tail. In most cases the intruder does not persist, but if he does, the female almost always swims away rapidly with both males pursuing her, jockeying for position and lashing their tails (I term this a "chase"). Eventually one male, most often the intruder, gives up and seeks another female. In some cases, both males seem to lose track of the female, and neither one can be considered the "winner."

These "chases" (Houde 1988b), in which two males pursue the same female, do not involve overt aggression between the males. Instead, each male appears to be trying to remain with the female and to exclude the other male. Chases with two or more males in pursuit of the same female are fairly frequent when the female is giving birth to young or has given birth recently. This may be because males are attracted by a female pheromone at this time (Crow and Liley 1979; Meyer and Liley 1982).

Direct aggression between individuals is seen much more rarely. A fish may dart toward another individual and appear to nip at it. The attacked individual usually springs away with fins folded. This kind of aggression occurs between two males, between two females, between two juveniles, or between an adult and a juvenile. Aggression between male and female is usually directed by the female to the male. Indeed, when male and female pairs are housed together with no other guppies, the female is very likely to become very aggressive toward the male and attack him repeatedly. This has the undesirable consequence of conditioning the male against all advances, even sexual responses, by females. Such males are reluctant to court females, and they flee when approached by a sexually responsive female. This can be inconvenient if the male is to be used in behavior observations.

More intense agonistic behavior sometimes occurs between males or between females, but, in my experience, this is not directly related to sexual competition. These interactions take the form of actual fights involving mutual darting and nipping, or formalized agonistic display. The agonistic display of guppies is termed the "lateral" display (Baerends et al. 1955, Liley 1966). The two individuals position themselves in parallel, most often pointing in the same direction, but occasionally nose-to-tail. They then spread their dorsal and caudal fins and quiver as in the sigmoid display, but broadside to one another. Males most often perform lateral displays, sometimes followed by fighting, while females most often fight immediately. During a lateral display, guppies become very dark and the color patterns of males tend to be obscured by this darkening. This is a further indication that the color pattern functions in intersexual display rather than in intrasexual competition or aggression. Other poeciliids have facultatively expressed color patterns that more clearly have evolved to function in an aggressive context (e.g., *Limia zonata*: Farr 1984; *Xiphophorus multilineatus*: Morris et al. 1995).

Aggressive behavior is rarely observed in natural social groups in the field or in the laboratory (Farr 1975). Aggression becomes more common when the sex ratio is artificially male biased, when there are very few individuals in an aquarium, or when a defensible food resource is present (Magurran and Seghers 1991). Aggression is particularly common in groupings involving only two males, even when there are two or more females. In these cases, one male almost always comes to dominate the other and attacks him frequently. Some other studies have also documented the formation of dominance relationships in laboratory situations (Gandolfi 1971; Gorlick 1976; Kodric-Brown 1992; Bruce and White 1995). Observations in the field (Farr 1975; pers. obs.) indicate that the levels of aggression seen in some laboratory situations are uncharacteristic of normal social behavior of guppies, and may even lead to pathological behavior of individuals. The fact that aggressive displays and dominance relationships have evolved in guppies suggests that they have some function even if they do not normally occur frequently. The most likely functions include sexual competition (see chapter 5) and competition for food (Magurran and Seghers 1991), but more study is needed.

2.6 Summary

Guppies are livebearing fishes with internal fertilization. As a result, matings are not associated with oviposition sites, but can occur at any place and any time. This means that male guppies approach females to court them rather than vice versa, and that the courtship activity of males is continuous.

Observations of the sexual behavior of guppies suggest that female choice is likely to be important in sexual selection in this species. For the most part, females are in control of whether or not they will mate with a given male and actively avoid unwanted copulations. Males perform courtship displays in which the color patterns are conspicuous to females, but they do not engage in overt aggression or direct combat in a sexual context. Most inseminations are the result of a sequence of behaviors requiring the female to cooperate with the male. These courtship sequences can be terminated at any time by the female or the male. The mating system of guppies is not determined entirely by female choice, however. Males do interact while courting females, and attempt sneak copulations with gonopodial thrusts not preceded by displays.

Males court many females in sequence, and usually perform several displays to a female before successfully copulating or giving up and courting a different female. A female, in turn, may or may not respond to the male's displays, and may or may not copulate with him. The distinctive gliding response of females and easily observed copulations leading to internal fertilization facilitate the study of sexual selection in guppies. For example, we can dissect the ways in which specific behavioral interactions between males and females result in sexual selection on male characteristics, via differential mating success. Chapter 3 examines the evidence for sexual selection in guppies and the roles of female choice and male-male competition.

3 Choosy Females and Competing Males: Mechanisms of Sexual Selection

3.1 Sexual Selection

In its original formulation, Darwin's (1859) idea of natural selection attempted to explain the evolution of characteristics of organisms through their adaptive value, primarily in terms of enhancing survival. For example, the streamlined flippers and body shape of marine mammals enhance their swimming efficiency relative to terrestrial ancestors and result in improved ability to pursue prey and escape from predators. In the same way, many of the characteristics of organisms and much of the diversity of life can be understood through relatively straightforward processes of natural selection and adaptation as originally set forth by Darwin (1859). But secondary sexual characters such as the elaborate plumes of birds-of-paradise and peacocks, elongated fins of fishes, the colors of guppies or the courtship displays of many species do not appear to have any survival-enhancing function and are not easily explained by the theory of natural selection in its original form. Instead, characters like these seem to be ornamental in nature, and probably involve physiological trade-offs (Partridge and Endler 1987; Gustafsson et al. 1995; Zuk et al. 1995a). If anything, elaborated plumes or fins are likely to reduce an individual's maneuverability and speed of flight or swimming (Møller 1989; Evans and Thomas 1992; Balmford et al. 1994; Garcia et al. 1994), making it an easier target for predators. Likewise, conspicuous color patterns or displays like those of guppies are likely to increase an animal's risk of being seen and captured by a predator (see chapter 1). To explain the evolution of secondary sexual characters requires an extension of the basic idea of natural selection leading to adaptation.

Darwin's (1871) simple solution is the idea of sexual selection—that secondary sexual characters could function to enhance the mating success of individuals. Sexual selection is a special case of natural selection, and the chain of logic is exactly parallel. Phenotypic variation in some character in one sex (usually males) leads to differential mating success: individuals with particular values or states of the character obtain more matings than individuals with different values or states of the character. If the variation in the character is heritable, then the differences in mating success can result in evolutionary change of the character from generation to generation. Darwin (1871) noted that either intersexual selection (mate choice) or intrasexual selection (competition within a sex) could be responsible for differences in mating success. Studies of sexual selection have sought to demonstrate patterns of differential mating success and to determine whether inter- or intrasexual selection is the mechanism involved.

It is important to note that the term "sexual selection" is most properly used to refer specifically to the process by which differential mating success can lead to evolution of secondary sexual characters. Behavioral processes resulting in sexual selection are properly termed "mate choice" or "mating preferences," not sexual selection. Finally, "sexual selection" is sometimes used to refer to the entire field of study related to sexual selection and its mechanisms, including the evolution of mate choice.

Although sexual selection would seem to be the most likely explanation for the evolution of color patterns, another possibility to consider is that the color patterns of guppies could actually be cryptic, functioning to reduce conspicuousness against the stream-bottom background (Endler 1978). Two facts argue against crypsis as the only function of color patterns in guppies. First, the expression of color patterns in wild guppies is limited to males: females have no colored spots, suggesting that the selective regimes affecting color patterns are different between males and females (Fisher 1930a; Haskins et al. 1961; Endler 1983). Sex limitation implies that color patterns are favored by some form of selection in males but are disadvantageous in females. It is unlikely that selection for crypsis would be so different between the sexes. Second, there is ample evidence that color patterns in guppies increase rather than reduce visibility to predators (see chapter 1). If natural selection by predators favors less conspicuous color patterns, then some other agent of selection, probably sexual selection, must favor conspicuousness to maintain guppy color patterns. To some degree, color patterns may have evolved to minimize conspicuousness to predators while at the same time maximizing conspicuousness to females (Endler 1987, 1991), but it is now clear that color patterns are not purely cryptic.

This chapter discusses the evidence that color patterns and other traits in guppies lead to differential mating success and thus sexual selection. It goes on to evaluate the contributions of female mate choice and male-male competition as mechanisms of sexual selection. Although male mate choice occurs in guppies and many other species (see Andersson 1994), it is unlikely to lead to differential mating success and sexual selection in species like guppies. Male mate choice is therefore discussed as an aspect of male mating strategy in chapter 4. A number of studies of sexual selection in guppies have documented differences in male mating success associated with differences in color patterns and other morphological and behavioral traits of males, while others have looked specifically at female preferences and male-male competition affect selection on male traits.

3.2 Differential Mating Success

A variety of experimental results provide convincing evidence that conspicuous color patterns are maintained in guppy populations by a mating advantage that balances the increased risk of predation (Haskins et al. 1961; Endler 1980, 1983; table 3.1). Measurements of mating success in guppies are laborious, however, consume a great deal of space and time, and are difficult to generalize to natural populations. Although we know that sexual selection occurs in guppies, we have very little idea how strong it might be, or whether it is constant or variable in space and time.

To document sexual selection (differential mating success), we need to show that variation in mating success is related to variation in a character such as the male color pattern in guppies. Because copulation is brief, fertilization is internal, and young are born live (chapter 2), differential mating success is somewhat more difficult to document in guppies than in other species used in sexual selection studies. In territorial birds, as well as in insects with a long copulation duration, sexual selection can be inferred easily by comparing mean trait values of mated and unmated males (e.g., Borgia 1981, 1982; Hieber and Cohen 1983; Hill 1990). In territorial fishes such as sticklebacks, sculpins, and cichlids, sexual selection can be inferred easily by observing the actual spawning success of males in choice tests (e.g., Semler 1971; Downhower and Brown 1980; Noonan 1983). In guppies, on the other hand, copulations are more difficult to observe because they are brief and occur unpredictably. Alternatively, mating success can be estimated by inferring paternity from offspring phenotypes (see Appendix).

Consistent differences in male mating success associated with different color patterns have been documented in a number of studies. Haskins et al. (1961) conducted competitive breeding tests in which males from two different color pattern strains competed for matings with females. By using known Y-linked strains, it was possible to determine the paternity of offspring born to the females in the experimental groups by comparing the color patterns of sons and putative fathers. This method of measuring mating success has been used in subsequent experiments (Farr 1980a; Bischoff et al. 1985; Houde 1988b; Reynolds and Gross 1992; Kodric-Brown 1993), and could prove worthwhile in future studies. Haskins et al. (1961) judged

Table 3.1

Overview of Studies Investigating Sexual Selection on Male Traits in Guppies

A. Morphological Traits				
Origin of Guppies	Evidence for Selection	Significant Traits	Nonsignificant Elements	References
Lab strains	Mating success	Subjective "brightness"		Haskins et. al. 1961
Lab strains (Haskins)	Mating success		Subjective "brightness"	Farr 1980b
Mixed wild populations	Mating success, female response	Number of red or blue spots		Endler 1983
Lab (Haskins) and domestic strains	Dichotomous-choice test	Carotenoid area, iridescent area	Black, white, blue-green, total color area	Kodric-Brown 1985
New Mexico hotspring	Dichotomous-choice test	Carotenoid brightness (diet manipulation)		Kodric-Brown 1989
New Mexico hotspring	Mating success, side preference	Carotenoid, iridescent area	Black area	Kodric-Brown 1993
New Mexico hotspring	Dichotomous-choice test	Area and number of carotenoid and total spots	Area and number of black spots, area of structural color	Nicoletto 1993
Feral South African population	Female response	Orange area, black spots, gonopodium length	Black, iridescent, yellow, area, body area and size, dorsal fin length	Brooks and Caithness 1995b,c,d
Paria River, Trinidad	Three measures of female response	Carotenoid area		Houde 1987
Paria River, Trinidad	Female response	Carotenoid brightness (parasite manipulation)		Houde and Torio 1992
Trinidad streams: low predation	Dichotomous-choice test	Total pigmented area ("brightness")		Stoner and Breden 1988
Trinidad streams: high predation	Dichotomous-choice test		Total pigmented area ("brightness")	Stoner and Breden 1988
Trinidad populations	Female response	Orange in most populations, other elements variable	Variable	Endler and Houde 199[illegible]
Domesticated	Female response, mating success	Tail length		Bischoff et al. 1985
Lower Quare River, Trinidad	Female response	Body size, tail length	Carotenoid, black, iridescent areas	Reynolds 1993

that the color pattern of their *Pauper* strain was among the dullest and that the *Maculatus* and *Armatus* strains had color patterns much brighter to the human eye. In the competitive breeding experiments, pitting *Pauper* against *Armatus* and *Pauper* against *Maculatus*, *Pauper* produced consistently fewer offspring than either of the other strains.

Another experiment (Haskins et al. 1961) involved an unusual individual that was phenotypically male but had a typically female XX sex chromosome configuration and was nearly colorless. The daughters of the XX

Table 3.1 (*cont.*)

B. Behavioral Traits

Origin of Guppies	Evidence for Selection	Significant Traits	Nonsignificant Elements	References
Lab strains (Haskins)	Mating success	Display rate (positive effect), aggression (negative effect)		Farr 1980b
Lower Aripo River, Trinidad	Mating success	Display rate (negative effect)		Houde 1988b
New Mexico hotspring	Mating success	Display rate, dominance (=aggression)		Kodric-Brown 1992, 1993
New Mexico hotspring	Dichotomous-choice test	Display rate		Nicoletto 1993
Trinidad streams: low predation	Dichotomous-choice test	Male "responsiveness" (courtship activity)		Stoner and Breden 1988
Trinidad streams: high predation	Dichotomous-choice test		Male "responsiveness" (courtship activity)	Stoner and Breden 1988
Oxford University greenhouse	Dichotomous-choice test	Display rate (parasite manipulation)		Kennedy et al. 1987
Feral South African population	Female response		Display rate, thrust rate, other behaviors	Brooks 1996b

male could be identified by a yellow tail on treatment with testosterone, so the mating success of the XX male could be compared with that of a normal male in a competitive breeding test. Again, the more conspicuous male sired more than half of the offspring. These results confirm that sexual selection was occurring between the strains being compared. The association between differences in mating success and color pattern differences is consistent with the idea that bright color patterns are favored by sexual selection. This association is far from conclusive, though, because Haskins et al. (1961) compared mating success between only a few color pattern strains, and these strains were inbred lines that could have differed in mating success as a result of random fixation of alleles at loci unrelated to color pattern. The experiments of Haskins et al. (1961) did not distinguish between female choice and male-male competition, although they were of the opinion that competition among males was more important.

A later study by Farr (1980b) produced results that were in part consistent with those of Haskins et al. (1961). Farr (1980b) performed competitive breeding experiments with descendants of some of the same strains studied by Haskins et al. (1961). In experiments with groups consisting of two males each from a different strain and two females, the *Pauper* strain performed better than the *Armatus* strain and the *Maculatus* strain sired only slightly more broods than the *Pauper* strain. But in tests with groups

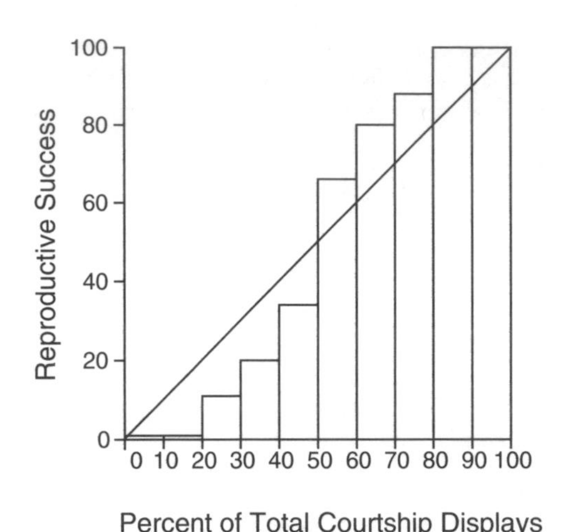

Figure 3.1 Effect of relative courtship rate on mating success in competitive mating trials. (Adapted from Farr 1980.)

consisting of ten males and ten females, the results were more consistent with those of Haskins et al. (1961). When *Maculatus* males were competing directly with *Pauper* males (five males per strain in the groups), they sired nearly all of the broods. In groups where the ten males included two males from each of five strains, *Maculatus* and *Armatus* were most successful and *Pauper* was least successful, again similar to the earlier results. Oddly and ironically, though, of all seven strains that Farr (1980b) studied, human judges found the pairs studied by Haskins et al. (1961; *Pauper* vs. *Maculatus* and *Pauper* vs. *Armatus*) to be the least different in consciousness rather than most different as assumed in the earlier study. Although this work provided some indication that differences in mating success might be related to color pattern, Farr (1980b) also found that the strains differed in the rate of sigmoid display per unit time. Males with high rates of display obtained disproportionately more matings than males with low rates of display (fig. 3.1), and Farr concluded that differences in coloration did not affect mating success independently of differences in courting rate. Given that the differences in behavior could have been the result of random fixation of alleles in inbred strains, it is difficult to generalize these results to natural populations.

In one of my own experiments, I was able to compare the mating success of "strains" of guppies with different Y-linked color patterns, but which had not been subjected to generations of inbreeding (Houde 1988b). Each "strain" consisted of sons and grandsons of twelve different wild-caught males. The related males in each strain shared the same color pattern (see fig. 1.1 for illustrations of the twelve color patterns). The relative mating success of the strains could then be estimated by examining the color patterns of offspring from competitive mating experiments exactly as in the Haskins (1961) and Farr (1980b) studies. I estimated the mating success of the males by counting the number of females from each group that had sons with a given color pattern. There were significant differences in mating success among the strains: some strains had consistently low success, while others had consistently high success.

These results are consistent with the results of Haskins et al. (1961) and Farr (1980b) in that genetic differences among color pattern strains do seem to be related to differences in mating success, even in "strains" recently derived from the wild. However, I found no obvious relationship between color pattern characters and mating success in this study—not surprising given that only twelve different color patterns were involved. Even though the strains were not inbred, there could still have been heritable differences in behavior or other characteristics that could account for the mating success differences. There was also no relationship between body size and mating success. Indeed, some of the most successful males were among the smallest.

All of these studies of mating success suffer from the problem of not comparing enough different color patterns to reach conclusions about the particular features of color patterns that are subject to sexual selection. The large amount of space needed to rear enough guppies to estimate mating success accurately is a constraint on the number of different males that can be compared in experiments like these. Differences in mating success among strains suggests that different color patterns may differ in mating success—this is evidence of sexual selection broadly speaking—but the data give no information about the specific characters under selection or the mechanisms of selection.

Endler (1983) was the first to demonstrate sexual selection for specific characteristics of color patterns and also the first to implicate female choice with quantitative data. His experiments involved direct observation of copulations and sexual responses of females. Endler's initial observation from long-term experiments in artificial streams (Endler 1980) was that populations free from predation evolved increased conspicuousness in that color patterns through a mismatch with the gravel background. Populations maintained on a coarse-gravel background evolved smaller spots while populations maintained on fine gravel evolved larger spots. Endler's (1980) interpretation was that males are most conspicuous to females when their spots differ most from the grain size of the visual background.

Endler (1983) tested this idea with actual observations of the behavior of females. Accordingly, he asked not only whether particular features of the male color pattern affected mating success, but also if the visual background affected relative mating success through effects on relative conspicuousness. The results confirmed both predictions (fig. 3.2).

The design of Endler's (1983) experiment was to place a single virgin

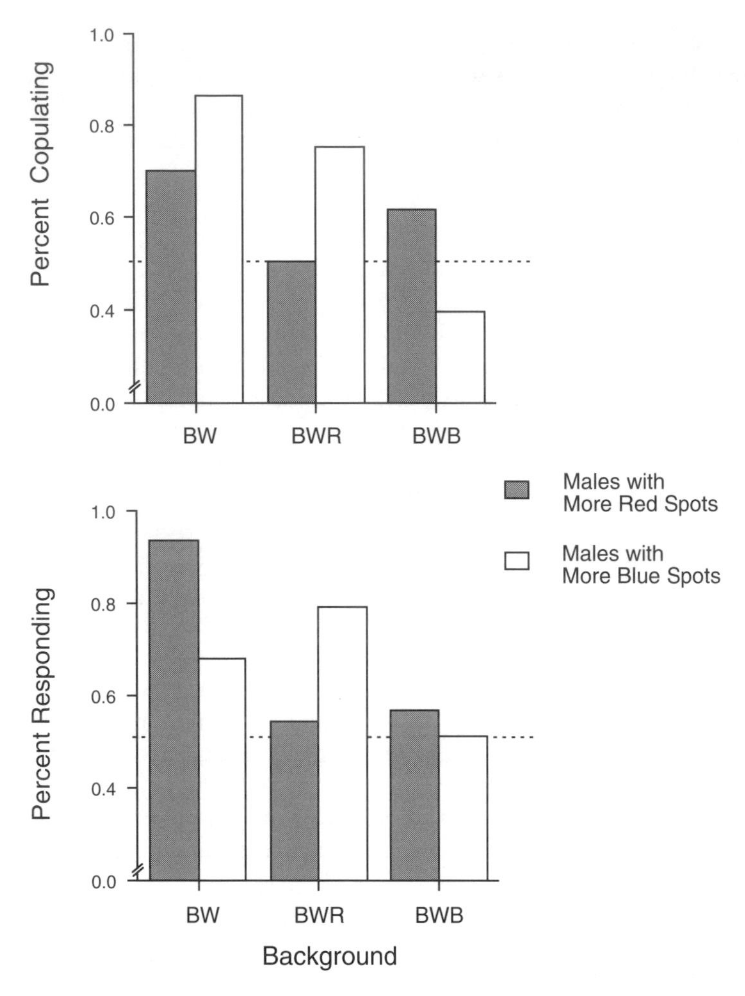

Figure 3.2 Effect of color pattern in relation to background gravel color on mating success of males in two-male, one-female competitive mating trials. Bars indicate fraction of males obtaining copulations (*top*) or female responses (*bottom*) for the male with more red spots (shaded bars) or more blue spots (open bars). Trials were conducted with black and white (BW), black, white, and red (BWR), or black, white and blue (BWB) gravel backgrounds. (Adapted from Endler 1983.)

female with two males and to observe which male elicited a sexual response from her and which was able to copulate with her. The males were ranked according to total number of spots, number of carotenoid (red-orange-yellow = "red") spots, and number of spots with structural colors (blue, green, silver, iridescent = "blue"). The trials were then conducted on different gravels, under natural light in a greenhouse. The males with the greater number of red, blue, or total spots had greater mating success than their competitors in trials on black-and-white and multicolored gravel (fig.

3.2). But results differed between trials with black, white, and red gravel and trials with black, white, and blue gravel. Males with more red spots had the greatest advantage when there was blue but not red in the background, and conversely, males with more blue spots had the greatest advantage when there was red but not blue in the background. The implication, of course, is that spots that contrast with the background have the advantage in sexual selection because they are most conspicuous to females (guppies have color vision; Endler, pers. comm.).

Another intriguing trend in Endler's (1983) experiment is that, with black-and-white or multicolored gravel, the advantage of males with more red spots is significantly greater than the advantage of males with more blue spots in the data on the initial sexual response of females. However, the reverse is true for the actual mating success of males. To put it more simply, having more red spots seems to be important for catching the female's attention, but given that a male has more red spots, then having more blue spots is important for keeping the female's attention long enough for successful copulation. This experiment was the first to use the behavior of females to infer mating preferences. A few subsequent studies have also measured both male mating success and female preferences simultaneously (Bischoff et al. 1985; Kodric-Brown 1989, 1993; see sec. 3.3 for details).

Endler (1983) provided some speculations about the adaptive significance of these patterns of female response and male mating success. He was not able to rule out the possibility that male-male competition affected the results of this experiment, though. It could be that the 2:1 sex ratio in his experimental groupings resulted in a stronger than normal effect of male-male competition on mating success, even though the data on female sexual responses may reflect actual mating preferences. This could explain the difference in results for female response compared with male mating success. More recent experiments have further tested the specific effect of color pattern differences on female preferences (see sec. 3.3), but more information is still needed on the determinants of actual mating success.

3.3 Female Choice

Mechanisms of Sexual Selection: Intersexual Versus Intrasexual Selection

Given that the color pattern of guppies is a good candidate for a sexually selected character, the next question to ask is, by which mechanism of sexual selection does the color pattern affect mating success? The studies described above demonstrate evidence for sexual selection, but data on male mating success alone does not discriminate between female choice and male-male competition as mechanisms of sexual selection. Darwin's (1871) two mechanisms of sexual selection, intersexual selection and intrasexual selection, may both play a role in guppies, but intersexual selection appears to be more important.

In intrasexual selection, members of one sex compete among one another for matings with members of the other sex. This usually means that males fight or contest with each other for access to females. Characters that may enhance success in male-male competition include large size or weapons such as horns and antlers (Conner 1988; Howard 1988; LeBoeuf and Reiter 1988). The guppy system is clearly not one in which males engage in direct combat (see chapter 1), but more subtle forms of male-male competition may occur (see below).

In intersexual selection, members of one sex exhibit preferences in their choice of mates from among the opposite sex. A character can evolve through intersexual selection if it improves an individual's chance of being chosen as a mate by a member of the opposite sex. In most species, females are the choosing sex and characteristics of males are subject to sexual selection, although there are exceptions in which sex roles are reversed (e.g., pipefishes, *Syngnathus typhle*; Rosenqvist 1990; see Andersson 1994 for other examples). Female choice has always been more contentious and more difficult to observe than male-male competition. Initially, people found it hard to accept the idea that female animals might be sophisticated enough to express what seem to be essentially aesthetic preferences. Eventually it became clear that females could exercise preferences, and that mate choice could be adaptive, especially if females could obtain immediate material benefits as a result (e.g., Yasukawa 1981; Nakatsuru and Kramer 1982; Gwynne 1984; Hill 1991). Female choice is thus relatively easy to understand and document in species with resource-based mating systems in which females obtain access to a territory, food for themselves or their young, or other forms of paternal care from the males they choose to mate with. On the other hand, the idea of female choice and the question of why it evolves has remained especially controversial in species in which, like guppies, males provide nothing but sperm to females before or after copulation.

Evidence for Female Choice

From first principles, the guppy system seemed promising for studies of female choice. Male guppies court one female after another and rarely interact or compete with other males. Generally, if a male is to be successful in mating, females must respond to his displays and cooperate in copulations (see details of sexual behavior in chapter 2). Therefore, female guppies are in a position to exercise mating preferences. Conspicuous color patterns can function via female choice in species recognition (e.g., Lorenz 1962) or mate attraction (see review in Andersson 1994), or may function in male-male competition (see sec. 3.4). Even casual observation of the courtship behavior of guppies suggests that male color patterns may function in attracting females. Male guppies are quite obviously flaunting their colors as they quiver and vibrate in front of females in their courtship displays (see chapter 2).

The color patterns of most Trinidad guppies probably do not play a role in species recognition, however. Areas of sympatry with closely related species are limited, and the high degree of polymorphism in color patterns would not seem to make them efficient signals for species recognition. Furthermore, males discriminate between conspecific and heterospecific females in interspecific choice experiments with guppies (Haskins and Haskins 1950), but not vice versa as would be expected if the color pattern were important in species recognition. On the other hand, Endler collected a distinctive guppy population from Venezuela that, at least in laboratory stocks, appeared to be monomorphic, but is interfertile with Trinidad guppy populations (pers. obs.). A species-recognition function for color patterns thus cannot be ruled out for all guppy populations. Studies of mate choice in Trinidad guppies described here generally assume that the color patterns function in intraspecific rather than interspecific discrimination by females.

Guppies have proved to be particularly useful for testing the role of female choice directly because of the individual variation in male color patterns and the ease of observation of female responses to males. The gliding sexual response behavior of females (Liley 1966; see also chapter 2) is an indication of the degree of attraction of a female to a particular male, and has been used to investigate female choice in a number of studies. Various other behaviors have also been used to assay female mating preferences, including mating speed (latency), aspects of the sequence of sexual responses leading up to copulation (chapter 2), and copulation itself. Many experiments with guppies and other species of fish and birds have presented females with a choice of males restrained by tethers or partitions and use the time each female spends in association with each male as a measure of female preference ("dichotomous-choice design"; see Appendix). The various measures of preference differ in how directly they may predict actual mate choice and the mating success of males and also in how likely male-male competition is to have confounding effects (see Appendix and descriptions of experiments below for details).

To demonstrate that differential mating success results from female choice, we must show that differences in the male character affect how females behave toward males, and that the differences in female behavior affect male mating success. Similarly, to demonstrate that male-male competition contributes to sexual selection, we have to show that variation in the male character specifically affects the success of males in their interactions with other males or in their ability to encounter and monopolize females. Experiments on female choice in guppies have the common goal of determining whether differences in color patterns or other characters of males are associated with differences in the behavioral responses of females. Most studies of female choice in guppies are based on observations of female behavior and make the assumption that this behavior reflects actual mating preferences and predicts male mating success. A few studies have measured effects of male characteristics on female behavior and male mating success simultaneously.

Preference for Area of Carotenoid Pigment

A number of different studies have documented that variation in the color patterns of male guppies affects the mating preferences of females. The strength of the evidence lies especially in the fact that very different experimental designs, in different laboratories, with different stocks of fish have all produced evidence for female choice based on color patterns. Single experiments taken alone generally have weaknesses or ambiguities, but together the evidence is convincing, even though results are not always consistent between studies. Table 3.1 lists the color pattern elements and other characteristics of male guppies that have been found to have significant and nonsignificant effects on female choice in a number of studies. The most frequently documented preference is that for carotenoid, or red-orange, color pattern elements.

Carotenoid colors are important in mate choice in many species and are of special interest in sexual selection studies because they are likely to reflect the nutritional and health status of the individual male. The role of such condition dependence in the evolution of female mating preferences is discussed in chapter 6. Threespine sticklebacks (McPhail 1969; Semler 1971; McLennan and McPhail 1989a,b; Milinski and Bakker 1990; Bakker 1993; Bakker and Mundwiler 1994), red jungle fowl, *Gallus gallus* (Zuk et al. 1990a,b,c, 1995a,b) and house finches (Hill 1990, 1991, 1992, 1993; Hill and Montgomerie 1994; Hill et al. 1994) are especially well studied examples of species in which mate choice is based on carotenoid characters. Other examples of fish with conspicuous carotenoid colors in male color patterns include North American darters, Etheostomidae (Lee et al. 1980; Kuehne and Barbour 1983; Pyron 1995) and African killifish, Cyprinodontidae (Haas 1976; Brosset and Lachaise 1995). Both the darters and the killifish include numerous species with bright male coloration and should be interesting systems for comparative studies of sexual selection on carotenoid colors and other traits.

Kodric-Brown (1985) studied the relative importance of several color pattern characters in males in female choice and found that carotenoid colors (red-orange-yellow) were most important. She used a dichotomous choice design (see Appendix) to test the preferences of each of twenty females for all possible pairings of eleven different males. This experiment tested not only the effects of different color pattern characters but examined the unanimity of choice of the different females. The males tested included three from fancy-tailed domestic strains, one each from the Haskinses' *Maculatus* and *Armatus* strains, and six different males from a "common" or domestic wild-type stock. Females included seven common stock females, seven *Maculatus* females, and six *Armatus* females. Kodric-Brown ranked males according to the relative areas of black, red-yellow, blue-green, iridescent, and white spots, and for total colored spot area. The color pattern rankings were then compared to the rankings by female preference measured in dichotomous-choice tests (fig. 3.3). Based on this correlation analysis, females appeared to prefer males with more red-yellow pigment area and with more iridescent area. Interestingly, six of the seven most preferred males in the study were from the common, wild-type stock. These had some of the largest amounts of carotenoid coloration. The preference for carotenoids of some females persisted when the iridescent area was statistically controlled for, but there was no preference for iridescent spots when carotenoid spots were controlled for. This is consistent with Endler's (1983) finding that carotenoid spots seemed to be important for initially attracting females.

Even though the females were from three different strains in this study, all but one individual showed close agreement in their rankings of the males. Kodric-Brown (1985) used this consistency of choice to argue for a common, adaptive basis for mate choice based on carotenoids, even in genetically diverse stocks. This is corroborated by studies showing preferences for carotenoid colors in other wild and feral populations (Endler and Houde 1995; Brooks and Caithness 1995a,b,c,d). Other experiments, however, have been able to detect significant variation in preferences among individuals and among populations (see chapter 6).

In a series of experiments (Houde 1987), I was able to demonstrate that mating preferences of females were based on variation in carotenoid coloration within a natural population of guppies. In designing these experiments, I was careful to use social groupings and encounters between individuals that mimicked the social structure of wild populations so that my results could be reasonably generalized to the natural situation. After many hours of observing guppies in the lab and in the field, I feel confident that their behavior in aquaria is representative of behavior in the wild, given that we recognize that differences in density, sex ratio, lighting, the presence of predators, and other factors (see chapters 4 and 5) can have impor-

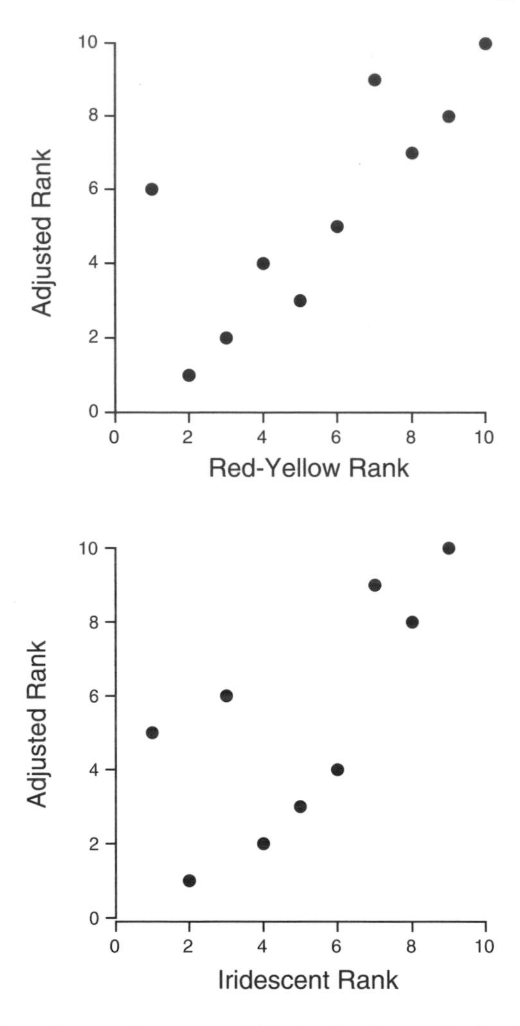

Figure 3.3 Female preference for carotenoid (red-yellow) spots (*top*) and iridescent spots (*bottom*) in male color patterns. Each male's ranked area of carotenoid or iridescent spots (X-axis) is plotted against his ranked attractiveness to females in dichotomous choice tests. (From data in Kodric-Brown 1985.)

tant effects. I used fish derived from the Paria River in Trinidad in my initial experiments because I had heard that they had extreme development of orange spots in their color patterns. This fact, combined with previous results on carotenoids (Endler 1983; Kodric-Brown 1985) and my own preliminary observations, led me to focus on the relative area of orange pigment as a possible cue for female choice.

I designed three experiments to test whether the dramatic variation in orange area in color patterns of Paria males affected female mating preferences (Houde 1987). Each used a different aspect of female behavior in response to males that could reflect the female's preference in mate choice. The first experiment examined the effect of orange area on the latency to copulation—mating speed—of a virgin female presented to a single male in no-choice trials. The most orange male was almost always quickest to mate, while the least orange males tended to be either slowest to mate or did not mate at all within the allotted fifteen-minute observation period. This suggests that females were in some sense more willing to mate, and to mate quickly, with males that had more orange in their color patterns. This could translate into effects on mating success in that males that mate more quickly might get more matings per unit time.

Like Kodric-Brown's (1985) experiment and other dichotomous choice experiments, the mating speed experiment eliminated possible confounding effects of male-male competition. In this case, male-male competition could be ruled out because males were tested individually, and the results can be attributed entirely to female choice. In addition, because females saw and responded to only one male and could not make comparisons, the observed differences in mating speed must reflect intrinsic or absolute preferences on the part of the female. More recent experiments have shown that female mating preferences can be modified depending on previous experience with males (Dugatkin 1992a, 1996; Dugatkin and Godin 1992b, 1993; Rosenqvist and Houde, in press; Breden et al. 1995; see chapter 6).

In the second and third experiments, I observed the responses of females interacting in groups with males, and also found that the orange area affected the responsiveness of females to male displays. In the group situation, females had the opportunity to compare males, and male-male interactions were possible and could have affected the results. In the second experiment, I observed the effect of variation in the orange area on the responsiveness of females to the males' courtship displays. I scored the attractiveness of a given male using the relative responsiveness of females to his displays: the fraction of displays that elicited a sexual response (see chapter 2) from females. The combined results of several trials (fig. 3.4) showed that females were most responsive to males with large orange areas and tend to be unresponsive to males lacking in orange. There is a significant leveling-off of the relationship in fig. 3.4, indicating that males with very large amounts of orange are no better at attracting females than those with moderately large amounts of orange. It is not clear if females were actively discriminating against the most orange males or if the leveling-off represents a psychophysical constraint on how females perceive and respond to males. Again, differences in the responsiveness of females to males depended on orange area and could have effects on mating success. The initial gliding sexual response of females is an essential part of the courtship sequence leading up to copulation, so that differences

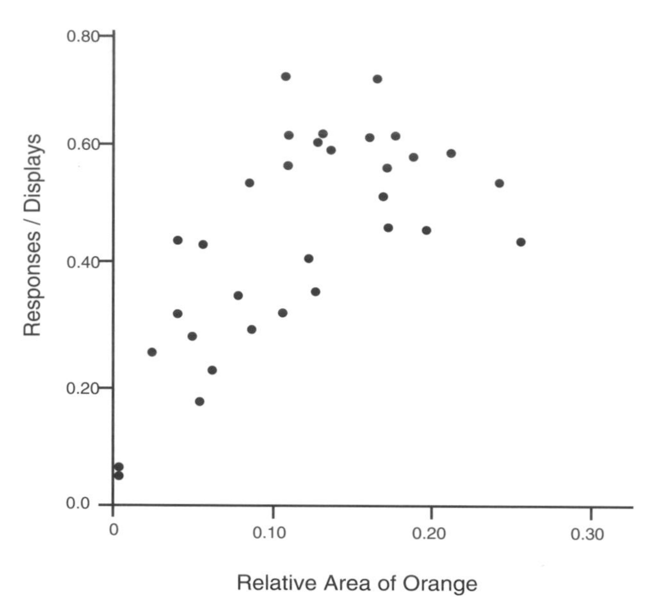

Figure 3.4 Effect of relative area of carotenoid (orange) spots on fraction of male displays eliciting a female response. (Adapted from Houde 1987.)

in the ability of males to elicit responses should affect their frequency of copulation.

The experiment described above showed that orange in the color pattern affects the effectiveness of a male's displays, but does orange have any effect other than in the context of courtship displays? My third experiment resulted from the serendipitous observation that virgin females sometimes perform the gliding sexual response to males that are just swimming past but are not displaying. Often, males are more interested in larger non-virgin females and appear not to notice small virgins as they swim by, but the small virgins often move toward the passing male with a characteristic gliding sexual response. I used the frequency with which males elicited glide responses from small virgin females in this context ("move-toward" behavior) as a measure of attractiveness, and to ask if orange area affected attractiveness even when males were not displaying. As in the other experiments, the small virgin females appeared to find males with more orange more attractive, even when the males were not displaying (fig. 3.5). This adds further support for female choice as a mechanism of sexual selection for color patterns in guppies, in combination with the other experiments and those of other researchers. The move-toward behavior itself could have effects on male mating success if it occurs in natural populations (however, it is possible that rearing virgin females separately from males makes them unusually receptive, even to males that are not displaying). Move-toward

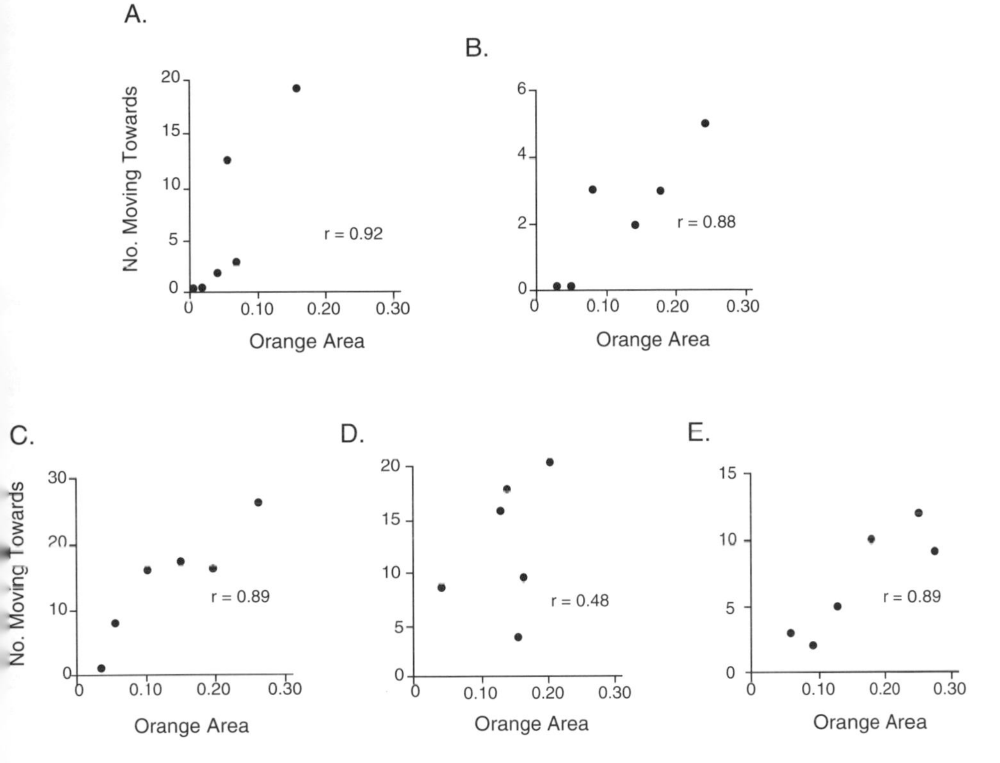

Figure 3.5 Effect of relative area of carotenoid (orange) spots on tendency of females to "move toward" nondisplaying males. Results from five different trials are shown in panels (A)–(E). (Adapted from Houde 1987.)

behavior usually occurs when the male appears not to notice the female initially but merely swims close by. Once the female performs the gliding move-toward response, though, the male often begins displaying to her. Males that inadvertently attract females this way may enhance their rate of encounter with receptive females and hence increase their mating success.

The experiments by Endler (1983), Kodric-Brown (1985), and Houde (1987) all indicate that female guppies use the relative area of orange in male color patterns as a cue in mate choice, and more recent studies continue to confirm these results (e.g., Brooks and Caithness 1995b,c; see table 3.1). A weakness of all of these experiments is that they examined the effects of existing variation in color patterns on female preferences. Without an experimental manipulation, we cannot rule out the possibility that some other character that is merely correlated with carotenoid area is the actual cue used by females. Courtship behavior (e.g., frequency of display) does not seem to be correlated with orange area in any of my experiments

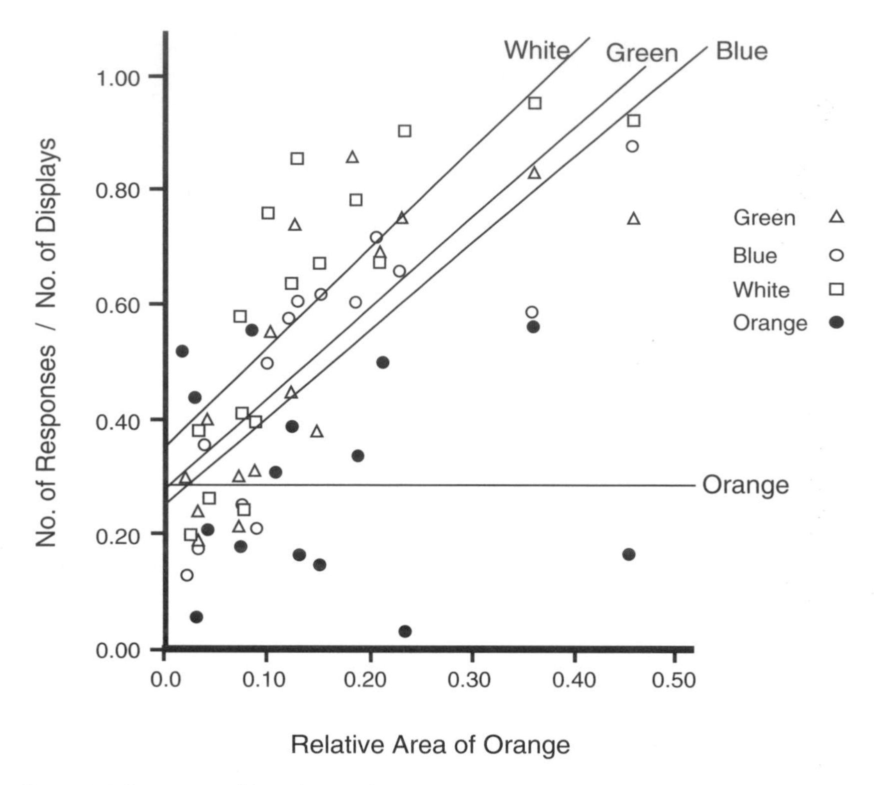

Figure 3.6 Response of female guppies to variation in orange area of male color patterns under white, green, blue, and orange light. Regression lines for each lighting condition are shown. Notice that regressions have significant slope for white, green, and blue light, but slope does not differ from zero for orange light. (Adapted from Long and Houde 1989.)

or in those of other researchers who have documented female choice. This still does not rule out the possibility that more subtle behavioral or morphological traits correlated with orange area could be the actual basis of choice.

Manipulating Females' Perception of Color Patterns

An experiment carried out by Kevin Long (Long and Houde 1989) addressed the question of whether females were responding to the color patterns themselves or to other correlated traits by manipulating the visual conditions under which female guppies view male color patterns. We repeated the experiment to test the effect of orange area on the responsiveness of females to displaying males (see section above), but ran replicate trials under white, blue, green, and orange light (fig. 3.6). Our reasoning was that different light colors should alter the perceived color and contrast of the spots, particularly orange spots, in male color patterns. Orange spots appear orange because they reflect orange light, and they contrast with the grayish body color of the fish, which reflects light from all wavelengths. Under orange light, orange spots continue to reflect orange as before, but the body color of the fish appears orange because the incident light is orange. The result is that the contrast of orange spots with the surrounding body color is reduced, and differences among fish that differ in orange area are less obvious. If orange spots are indeed the basis of female preferences, then preferences for more orange males should disappear under orange light.

Conversely, blue light is expected to increase the contrast of orange spots. But because blue is complementary to orange, orange spots reflect little or no light under this treatment and so appear black, or nearly so. We predicted that female preferences should persist under blue light if the contrast of the orange spots is important, but could disappear if preferences depend on differences in the amount of specifically orange wavelength light reflected by males. The problem with the last prediction is that the filter used to produce the blue light treatment would have to complement the orange spots exactly to completely eliminate reflectance. Our eyes were able to detect the orange color of the guppy spots under the blue light treatment, so this exact matching probably did not occur. Finally, the green light treatment served as another control in that the contrast of orange spots increased only slightly and remained distinctly orange in appearance.

The results of the light color experiment confirmed that the orange spots themselves are responsible for the preferences documented in other experiments (fig 3.6). If some correlated character unrelated to the color pattern had been responsible for the preference, then altering the visual environment should not have had any effect on preferences. Instead, the preference for males with more orange disappeared under orange light, consistent with the change in the appearance of the males. The preference persisted in the blue and green light treatments, though, as well as in the white light treatment. Because the blue filter was not perfectly effective in blocking orange light, we do not know if orange reflectance is essential for the female preferences or if the contrast of the orange spots with body color is the critical factor. Brooks and Caithness (1995c) replicated the light color experiment using feral South African guppies and obtained results similar to ours. In addition, they found that females tested under blue light showed a preference for the black spot area. Milinski and Bakker (1990) were also able to demonstrate the importance of red reflectance for female preferences in sticklebacks using a similar experimental method. In contrast with the guppy studies, they found that under green light complementary to the red throat color of the sticklebacks, females no longer showed the preference for intensity of red coloration that they showed under white light. This suggests that the red coloration itself, rather than the contrast of the red throat patch, is the basis for the preference in sticklebacks.

Female Preferences for Other Morphological Characters

The area of red-orange coloration is not the only sexually selected character of guppies. Other experiments have examined the effects of variation in other color pattern characters, tail and body size, and behavior of males. Mate choice in guppies appears to be based on many criteria, which can be teased apart statistically (Endler 1978, 1980, 1983, 1991, 1993; Kodric-Brown 1993; Brooks and Caithness 1995b; Endler and Houde 1995). Mate choice based on multiple criteria has been documented in a few other species, such as pigeons (Burley 1981), bowerbirds (Borgia 1985), and jungle fowl (Zuk et al. 1990b, 1995b), but is predicted by theory (Møller and Pomiankowski 1993b; Pomiankowski and Iwasa 1993; Iwasa and Pomiankowski 1994, 1995). We need to understand more about the relative importance of the different characters involved in mate choice and how they interact in order to obtain a complete picture of sexual selection in this system.

While the area of carotenoid spots is an obvious and easily quantified character, interest in the possibility of preferences based on condition-dependent characters has led to studies looking at variation in the brightness of orange coloration. Two studies have shown that female guppies prefer males with bright orange spots over males with dull orange spots. In one case, variation in brightness was the result of laboratory manipulation of diet (Kodric-Brown 1989), in the other, variation was due to laboratory manipulation of parasite infection (Houde and Torio 1992). See chapter 6 for further details of these experiments.

Other specific color pattern characters, such as the area of iridescent, black, or other spot types have been tested in various studies (table 3.1), but with little consistency of results. In a systematic survey of several guppy populations, Endler and Houde (1995) found female preferences for several different color pattern elements, but the preferences varied between populations (see details in chapter 6). Preference based on red-orange coloration appears to be the most consistent pattern of female choice among studies and among populations, with less pronounced preferences based on other color pattern elements. Genetic variation among populations may explain some of the inconsistency among studies in documenting preferences.

Although black spots are preferred by females in only a few populations (Endler and Houde 1995), black markings may affect mate choice through their interaction with orange coloration. The finding that females use black coloration as well as orange coloration in mate choice under blue light (Brooks and Caithness 1995c) suggests that black could be used as a "redundant signal" (Møller and Pomiankowski 1993b), especially when the primary cue, orange, is altered. A further study (Brooks and Caithness 1995d, Brooks 1996a) suggests instead that black spots may function to amplify (Hasson 1989, 1990, 1991) or improve detectability (Schluter and Price 1993) of orange spots rather than as redundant signals. See chapter 6 for further discussion of a possible role of amplifier traits.

Brooks and Caithness (1995c) used a freeze-branding technique to remove a black spot from experimental males while they freeze branded an unpigmented area on control males. Within each pair, control and experimental males were brothers and so had otherwise nearly identical color patterns. Even though natural variation in black coloration did not seem to affect female preferences in this population, removing the black spot reduced the attractiveness of experimental males. This suggested that female choice is important in maintaining black spots, and possibly other nonpreferred colors in guppy populations (Brooks and Caithness 1995c). However, further analysis of the data (Brooks 1996a) revealed that females preferred males with greater orange area before the freeze-branding treatment but that this preference disappeared afterwards. This suggests that black spots might function as a signal amplifier (Hasson 1989, 1990, 1991), enhancing the ability of females to discriminate among males based on orange coloration (Brooks 1996a). Alternatively, the change in black spot area might have been great enough to override differences in attractiveness due to natural differences in orange area.

The color patterns of guppies occur on both the right and left size of each individual, and are sometimes asymmetrical. Given recent interest in fluctuating asymmetry as a possible condition-dependent cue for mate choice (e.g., Møller and Pomiankowski 1993a; Watson and Thornhill 1994; see chapter 6 for further discussion), a number of studies have attempted to relate the symmetry of guppy color patterns between the left and right sides of males to female mating preferences. So far, there is no evidence for female preferences for symmetrical males in guppies (Brooks and Caithness 1995b; Nordell 1995; Endler and Houde, unpublished data).

Other poeciliid species exhibit a variety of color patterns, some of which, like those of guppies, are sex limited and polymorphic (Liley 1966; Farr 1984; Angus 1989). There is surprisingly little information about mate choice based on color patterns in species other than guppies. The black-spotted morph of *Gambusia holbrooki* has disproportionately high mating success, possibly attributable to female choice (Karplus and Algom 1996; but see Nelson and Planes 1993). On the other hand, different color morphs of the pygmy swordtail *Xiphophorus pygmaeus*, did not differ in their ability to attract females (Baer et al. 1995). Further information about sexual selection and female choice on color pattern elements in other poeciliid species is needed.

Females base their mating preferences on male size in numerous species (see Kodric-Brown 1990, table 1; Andersson 1994, table 6.A). Among poeciliid fishes, male body size is known to affect female choice in mosqui-

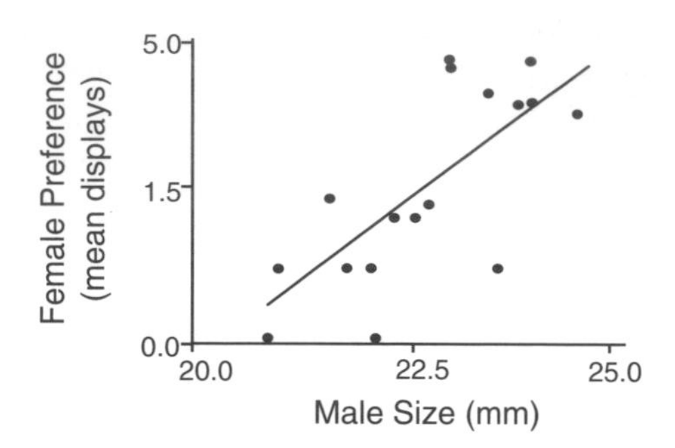

Figure 3.7 Effect of male size on female responsiveness. (Adapted from Reynolds and Gross 1992.)

tofish *Gambusia affinis* (Hughes 1985; McPeek 1992), Gila topminnows *Poeciliopsis occidentalis* (Constantz 1975), sailfin mollies *Poecilia latipinna* (Ptacek and Travis, unpublished results), and the swordtail *Xiphophorus nigrensis* (Ryan and Wagner 1987; Ryan et al. 1990c). Female three-spined sticklebacks also favor large males (Rowland 1989a,b). In most of these species, male body size variation is related to alternative mating tactics and male-male competition (see below).

Guppies differ from the species listed above in that male body size appears to be less important in all aspects of sexual selection (see further discussion below and in chapter 4). Reynolds and Gross (1992) reported female preferences for larger males (fig. 3.7) in a guppy population from the Lower Quare River of Trinidad. They observed the responses of single females interacting with single displaying males and found a correlation of the response rate of females with the total length and tail length of males. In data from a comparison of eleven populations, Endler and Houde (1995) found a preference for standard length (body length not including tail fin) in one population, a preference for tail fin size in two populations, a preference for total length in one population, and a preference for body height in two populations. Brooks and Caithness (1995b) found that gonopodium length also affected female responsiveness in a feral South African guppy population.

Bischoff et al. (1985) demonstrated a preference for males with long tails in dichotomous choice experiments involving domesticated "fancy-tailed" guppies. The long-tailed males also had greater mating success than short-tailed males when the fish were allowed to interact freely (fig. 3.8). The results of Bischoff et al.'s (1985) experiment are difficult to interpret, though, because the experimental fish were from artificially selected strains with more highly elaborated tails than any found in nature. It would be interesting to know if females from wild guppy stocks have any similar preference based on such extreme differences in tail length.

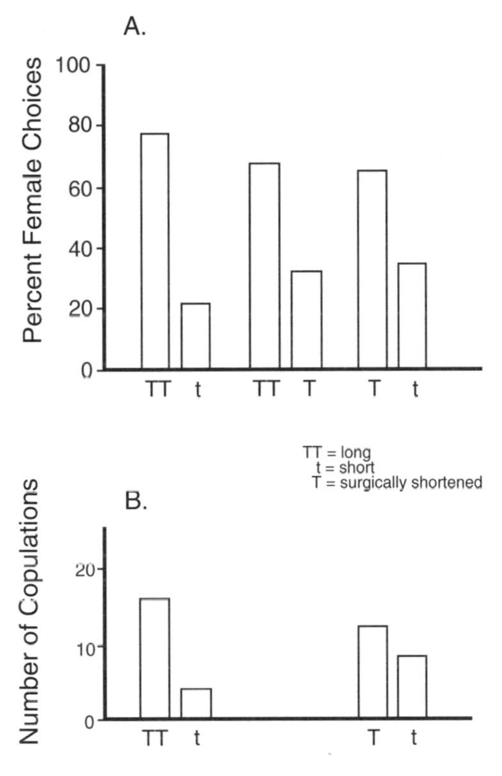

Figure 3.8 Effect of tail length on female preference and male mating success. (Adapted from Bischoff et al. 1985.)

Most of the variation in tail length in natural populations of guppies seems to be related to age. Old males sometimes develop elongations of upper, lower, or central portions of the caudal fin corresponding to areas of pigmentation on the tail (pers. obs.). These elongations can be as long as or even longer than the caudal fin itself! These age-dependent elongations in guppies are similar to elongations ("swords") of swordtails, but are probably not, strictly speaking, homologous characters (Basolo 1995b). Basolo (1990a,b, 1995a,b) has demonstrated female preferences for swords and for sword length in *Xiphophorus* species and a related genus, including species in which males lack swords. Haines and Gould (1994) have suggested that platyfishes *Xiphophorus variatus*, may have a sensory bias based on the ventral length of the tail leading to a preference both for ventral sword structures and for elongation of the whole tail. A similar bias might explain the preference for artificially selected long tails in Bischoff et al.'s (1985) study.

Various aspects of body and tail size thus appear to be important in female choice in some guppy populations, but this kind of preference is not found in all populations. Further experiments are needed in which body size is manipulated through diet or in which the effects of naturally occurring differences in tail length are examined and manipulated.

Female Choice for Rare Males?

Many of studies on sexual selection in guppies described here have addressed the question of why males have conspicuous color patterns, and most of these support the idea that color patterns affect male mating success, often through female choice. An equally fundamental question is why guppy color patterns show such high levels of genetic polymorphism. This question has been addressed directly by only one study of the rare-male effect in guppies (Farr 1977; see Partridge 1983 for review), and the results of that study need to be followed up and extended. Baer et al. (1995) found no evidence for a rare-male preference in female *Xiphophorus pygmaeus*. The idea that a color pattern polymorphism can be maintained by mate choice is supported by Borowsky and Kallman's (1976) finding that *Xiphophorus maculatus* mate disassortatively based on tail-spot patterns.

Farr (1977) tested the hypothesis that the polymorphism in color patterns is maintained by a rare-male mating advantage. He conducted competitive breeding tests in which ten virgin females were allowed to interact and mate with ten males. Of the ten males, nine were from the same strain (*Pauper*, *Maculatus*, or *Armatus*), and one was from a different strain. Farr then scored the paternity of broods produced by the females. Of a total of twenty-one broods counted, nine had been sired by the rare male. The mating success of the rare males was thus far greater than expected under random mating. Farr (1977) also noted that females showed a high rate of sexual responses to males that were newly introduced to an aquarium population. Based on these observations, he concluded that females may prefer males with rare phenotypes, and that this preference may be advantageous because it tends to produce fitter, more heterozygous offspring.

The rare-male effect and experiments purporting to provide evidence for it have been criticized on several points (e.g., Partridge 1983), and the idea has been largely ignored for a number of years. A weakness of Farr's (1977) study is that the competitive breeding experiment did not directly demonstrate that the females preferred rare males; the mating advantage could have been the result of male-male competition. Also, in guppies, there is so much polymorphism that most males in a given group are likely to have unique, and hence rare, color patterns. Nevertheless, given that the high levels of polymorphism remain unexplained, additional experiments examining the possibility of frequency dependence in female choice and male mating success could provide especially interesting results.

Female Choice and Male Display Behavior

Yet another kind of secondary sexual character of males that could be a basis for female choice is the courtship display itself. Courtship displays are clearly directed toward females and appear to attract receptive females. Display activity is likely to be affected by sexual selection in that the more displays a male performs, the more matings he is likely to obtain, regardless of female preferences. We would like to know whether display behavior is, in fact, used as a cue for female choice. Despite the conspicuousness of courtship displays in many species and their ease of observation, relatively few studies have shown that attributes of displays are used as a cue in mate choice in fishes (Andersson 1994, table 6.A; but see Knapp and Kovach 1991, Karino 1995, Grant and Green 1996, and Rosenthal et al. 1996). Female preference for courting over noncourting male morphs have been demonstrated in sailfin mollies (Ptacek and Travis, unpublished results) and swordtails (*X. nigrensis*: Ryan and Wagner 1987; Ryan et al. 1990c), but the differences in courtship are confounded with differences in body size in both species (Zimmerer and Kallman 1989; Travis 1994; but see Rosenthal et al. 1996).

Female choice for visual display behavior has been demonstrated in bird species such as sage grouse *Centrocercus urophasianus* (Gibson and Bradbury 1985), Lawes' parotia *Parotia lawesii* (Pruett-Jones and Pruett-Jones 1990), and Jackson's widowbird *Euplectes jacksoni* (S. Andersson 1989, 1991, 1992), and red jungle fowl (Zuk et al. 1990b, 1995b). Mate choice based on vocal displays in birds, insects, and anurans is more commonly demonstrated (Andersson 1994, table 6.A).

The lack of evidence for mate choice based on display in fishes may be related to the interdependence of male displays and female responses (see Liley 1966 and discussion below). If males attune their display behavior to the responses of females, then it is difficult to determine cause and effect. In the bird examples cited above, displays are used more in the context of long-range attraction, prior to the arrival of females, as are vocal displays. Displays by fish, on the other hand, are less likely to be visible over long distances and are more likely to function in interactions when males and females are in close proximity. A direct effect of display behavior on mate choice is thus more difficult to demonstrate in species like fishes.

There is evidence for mate choice based on display in several studies of guppies, but careful interpretation is needed. In observations of one-male, one-female pairs of guppies, Nicoletto (1993) found that the frequency of female sexual responses was positively correlated with the frequency of male displays. If every female has a fixed probability of responding per display, however, this relationship might not be considered active discrimination on the part of females, even though sexual selection on the male character could result. On the other hand, differences in display behavior could lead to active discrimination if females further modified their behavior upon encountering a male with a high display rate.

Several dichotomous-choice experiments have found that females spend more time near the more actively displaying male (Bischoff et al. 1985; Kennedy et al. 1987; Stoner and Breden 1988; Kodric-Brown 1993), suggesting a preference based on display rate. The dichotomous choice design is subject to difficulty in interpreting data on courtship rates, though. We can never be sure if a female was really attracted to a male because of his higher display rate or if the female's interest, perhaps in the male's color pattern, stimulated the male to display at a high rate. Studies of mate choice based on display behavior in all species, not just guppies, are subject to this difficulty of interpretation. The experimental design may also affect the way in which female behavior affects courtship by males. In experiments using freely interacting groups of guppies, courting rate and attractiveness are either not associated (Brooks 1996b) or are negatively associated (Houde 1988b; see details below this section).

Kennedy et al. (1987) compared the time females spent near males that varied in parasite burden (nematodes: *Camellanus cotti*) and found that females preferred the male with the lower burden. The more heavily parasitized male almost always had a lower courting rate than the less-parasitized male. Kennedy et al. (1987) suggested that females might be using courting rate as a cue for mate choice in this case. However, females could have been detecting some other effect of parasites and discriminating against heavily parasitized males by not associating with them, and this, rather than the parasites themselves, could have led to the difference in courting rates.

In a similar experiment, Stoner and Breden (1988) examined the effect of naturally occurring variation in courtship activity on female choice and found that females associated with the more active males. The females might actually have had a preference for the more frequently courting males, but alternatively, the brighter males may have courted more frequently as a side effect of being preferred by females. Stoner and Breden (1988) attempted to control for this difficulty by using a measure of the male's courtship activity per female visit, so that the amount of time the female spent with each male was factored out. Still, subtle differences in how females behave while they are near a given male could affect his courting behavior rather than vice versa. Bischoff et al. (1985) eliminated the possibility that male behavior depended on the responses of females by placing their males behind one-way glass so they could not see the females. They noted that the males seemed to display spontaneously even though they could not see the females, but in general male guppies do need to see a female before they will display.

Rosenthal et al. (1996) got around the usual difficulties of dichotomous choice tests in a study of *Xiphophorus helleri* by using video animation techniques. They presented simultaneous, paired images of the same male either displaying, feeding, or not moving to females. Females preferred the courting males over the other images. This method allowed effective separation of the effects of courtship behavior from the effects of size or other traits. In a similar experiment, Rowland (1995) was able to manipulate a specific attribute of male displays, the tempo of the zigzag dance, and determined that both male and female sticklebacks showed reduced attraction to very slow and very fast displays. This technique has great promise for investigating the effects of display behavior and other characters on mating preferences in guppies.

The display rate of males may affect mating success as a result of female preference for high display rates or possibly as a result of male-male competition. Farr's (1980b) finding that males with high rates of display obtained disproportionately more matings than males with low rates of display (see sec. 3.2 above and fig. 3.1) suggests that females prefer the more frequently courting males. An alternative explanation could be that males with higher display rates are more dominant and can perform more effective displays, suppress the displays of other males (Bruce and White 1995), or are interrupted less often than subordinate males. All of these could lead to disproportionately higher mating success of males with higher courting rates. Experiments by Kodric-Brown (1992, 1993) partly support this latter interpretation. In trials where two males were allowed to interact with each other and mate with one female, the more dominant (i.e., aggressive) male displayed at a higher rate than the subordinate male. However, when the males were separated in dichotomous-choice tests, display rates were not significantly different between dominant and subordinate (fig. 3.9), suggesting that the dominant individual had suppressed courtship by the subordinate individual when they were allowed to interact freely. The difference in display rate between dominant and subordinate males may account at least in part for the positive effect of display rate on mating success in Farr's and Kodric-Brown's studies. However, Kodric-Brown (1993) also provided more conclusive evidence that display rate affected female preferences independently of dominance in dichotomous-choice tests.

Finally, I found results contradictory to most other studies in my study of mating success (Houde 1988b). In group situations, male display rate to females was actually negatively correlated with mating success. The most attractive and successful males were those that elicited the greatest fraction of responses from females, but these males tended to have a reduced rate of

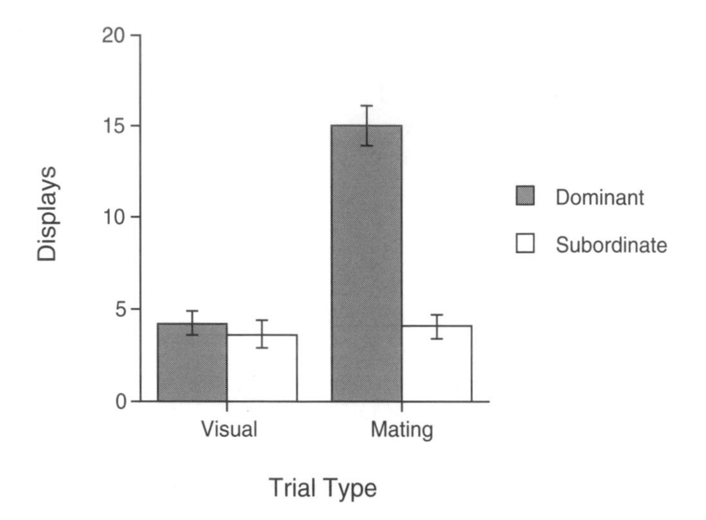

Figure 3.9 Effect of dominance on display rate in trials in which males were separated by barriers (visual) or freely interacting (mating). (From data in Kodric-Brown 1993.)

courtship. In this situation, I think that successful males were actually slowing their courtship and being more deliberate once females responded to them. They may have been shifting their use of display from a means of probing female receptivity to a means of stimulating a receptive female to cooperate in a copulation. This again suggests that courtship behavior of male guppies may depend on the behavior of females. There is thus a need for detailed, careful experimentation to tease apart the interdependent effects of male courting rate and female preferences.

Given that all of the studies cited above have examined the effects of courting rate, it would be interesting to examine the effects of other aspects of male displays on female choice. For example, the duration, pulse rate, and vertical displacement of sigmoid displays might all be more informative about male vigor than display frequency.

Female Preferences and Male Mating Success

Most of the best data on the mating preferences of female guppies involve observation of female behavior, which in turn is assumed to predict the actual choice of mates and, hence, male mating success. But actual matings are rarely observed to complete the chain of inference that female choice is a mechanism of sexual selection in guppies. The lack of a direct relationship between the behavioral preferences of females and differential mating success of males is a source of criticism of much of the empirical work on mate choice. Direct observations of female behavior do have the advantage that possible effects of male-male competition can be eliminated through experimental design, strengthening the case for female choice. On the other hand, actual mating success in natural populations may be influenced by male-male competition as well as by female choice so that inferences about the evolutionary effects of sexual selection must be made cautiously. What evidence is there to support the assumption that behavioral preferences of females result in sexual selection on male traits?

There are a number of studies in which behavioral preferences of female guppies for a specific trait and its effects on male mating success have been documented in the same study (Bischoff et al. 1985; Kodric-Brown 1989, 1993; Reynolds 1993; Brooks 1996b). These experiments provide evidence that female preferences do indeed result in sexual selection as an evolutionary process, and validate behavioral methods used to measure female mating preferences. Bischoff et al. (1985) documented a female preference for males with longer tails in a dichotomous-choice experiment and also showed that the males the females appeared to prefer were indeed more likely to copulate. Similarly, Kodric-Brown (1989) documented a preference and mating advantage for males fed a high carotenoid diet relative to males fed a diet lacking in carotenoids using dichotomous-choice tests followed by direct mating competition (more details in chapter 6). Kodric-Brown (1993) showed that coloration and display rate both affect association preference by females in dichotomous-choice tests and that the behavioral preferences predict the subsequent mating success of males. She estimated male mating success by determining the paternity of offspring by their color patterns (see Appendix). These studies lend credence to the inference that side association data from dichotomous-choice experiments really do reflect mating preferences and predict the direction of sexual selection.

The responsiveness of females to male displays in interacting groups of guppies (open aquarium experiments; see Appendix) also appears to predict differential mating success. I was able to show (Houde 1988b) that the differences the rate of glide responses by females to male displays (see chapter 2) was correlated with male mating success, though not very cleanly. I also estimated male mating success by determining the paternity of offspring by their color patterns (fig. 3.10), but no specific male trait appeared to predict mating success in this study. The fact that the predictive relationship between female preference and male mating success is not very clean in this experiment suggests that male-male competition or perhaps stochastic variation could play a role. Although females in this study did not seem to show preferences for particular male phenotypes, the important implication is that the glide response of females can be used to infer sexual selection by female choice in other studies. In a separate study, also using an open-aquarium design and assessing paternity using color pattern, Brooks (1996b) found that both the rate of initial orientation of females to the male and the rate of glide responses predicted male mating success.

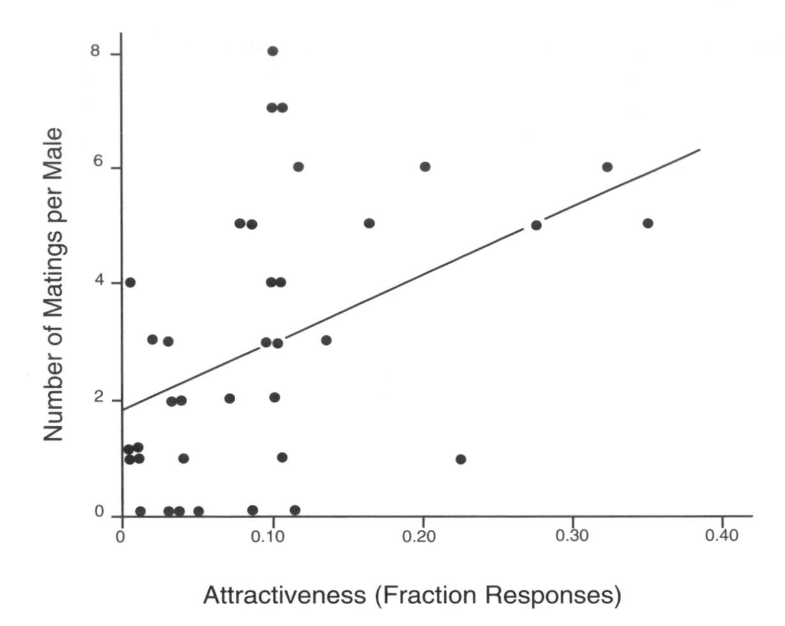

Figure 3.10 Relationship of male attractiveness (fraction of displays eliciting a female response) to mating success in a competitive mating experiment. (Adapted from Houde 1988b.)

These results suggest that the behavioral preferences of females for male traits found in other studies can reasonably be assumed to lead to sexual selection on male color patterns.

I attempted to test this inference more directly in an unpublished study by measuring the distribution of orange area in the offspring of females that had the opportunity to choose their mates compared with offspring of females that had no choice of mate. "Choice" females were allowed to mate in a large, freely interacting group (forty males and forty females in a 1.5-diameter wading pool) and then isolated and allowed to produce offspring, while "No Choice" females were randomly assigned to mate with one male. Given that orange area has very high heritability in the Paria population (chapter 1), I predicted that females allowed to exercise their preference should have offspring with a greater orange area than females that were not allowed to choose their mates. The results were anything but conclusive. There were slight differences in the predicted direction in four of five replicates, but virtually no difference in the means in the fifth replicate. A possible explanation is that females in the "Choice" treatment may have been undiscriminating in mating behavior because they had no prior experience with mature males (see discussion in chapter 2). Alternatively, the density of fish and physical homogeneity of the wading pool might have encouraged competitive interactions among males and prevented the females from choosing their mates freely. Finally, the sample sizes could have been too small to detect even a moderate effect of sexual selection. These results suggest that the behavioral preferences that we see may not always result in an evolutionary effect on the preferred male trait. An example of a guppy population in which color patterns do not seem to match mating preferences (Houde and Hankes, in press) is described in chapter 6. More work is needed to determine the extent to which behavioral preferences of females translate into sexual selection on male phenotypes and to understand factors that might prevent females from mating with males they prefer.

3.4 Does Male-Male Competition Play a Role?

Although there is good evidence now that female choice is a mechanism of sexual selection on color patterns and other characters in male guppies, the possible role of male-male competition is much less clear. Unlike territorial species such as sticklebacks (Bakker 1994; Rowland 1994), wrasses *Thalassoma bifasciatum* (Warner and Schultz 1992) and damselfishes *Stegastes nigricans* (Karino 1995), in which males compete among themselves for the territory itself and then attempt to attract females to that territory, male-male competition in guppies and most other poeciliids is more likely to affect access to females directly. Male-male competition and female choice are therefore likely to overlap in time and their effects are more difficult to separate though observations. In guppies, prior interactions may affect how males respond to one another while courting females, but this is a more indirect effect.

Male guppies are smaller than females, and they have no weapons. They are thus not obviously adapted for overt combat. Male-male competition could involve the use of color patterns in intimidation displays between males (e.g., LeCroy 1981; Rowland 1982, 1989b; Hansen and Rohwer 1986; Bakker and Sevenster 1983; Morris et al. 1995), but displays between male guppies are rare. This needs more study, however. For example, Kodric-Brown (1993) found that dominant individuals were more colorful than subordinate individuals, but the mechanism for this association is not clear. In natural social groupings of guppies (many males and many females), overt aggressive behavior between males is rare, especially in a sexual context (chapter 2). However, aggression does sometimes occur in experimental groupings in aquaria, especially when only two or three males are present (Ballin 1973; Gorlick 1976; Farr 1980a; Kodric-Brown 1992, 1993; Bruce and White 1995), so male-male interactions could affect mating success under some conditions. It is difficult to draw clear conclusions about male-male competition in studies of guppies because dominance relationships could be set up quickly, and prior interactions could have subtle effects on the ways males interact and on their mating success (Kodric-Brown 1992, 1993).

Some other poeciliid species have more obvious aggressive interactions among males (e.g., Beaugrand et al. 1984; Farr 1984, 1989; Bildsøe 1988; McPeek 1992; Morris et al. 1995), and male-male competition may well play a greater role. Bisazza (1993) compared the relative roles of female choice and male-male competition among poeciliid species and concluded that female choice is important in species like guppies in which males court females and females cooperate in copulations. Male-male competition (either indirect or direct) is more important in species lacking courtship in which males inseminate females without cooperation, and in species which defend territories. Our present understanding of the guppy mating system is consistent with this interpretation, although it is clear that both female choice and male-male competition play a role in this species. The ability of male guppies to obtain copulations both through courtship displays and through sneak copulation attempts not involving female cooperation is considered in chapter 4. Most of the comparative information about poeciliid mating systems reviewed by Farr (1989) and Bisazza (1993) is based primarily on patterns of occurrence of male behaviors, and conclusions about the relative roles of male-male competition and female choice are primarily inferential. There is relatively little information on how male-male interactions affect male mating success and contribute to sexual selection on particular characters. In fact, some of the best data on this comes from studies of guppies.

Male-male competition could operate in two ways in guppies that are distinct from the effects of female choice. First, males may compete indirectly (scramble competition), attempting to maximize their rate of encounter with responsive females, but without directly interfering with other males. Each copulation that one male obtains is a gain for him, but a loss for other males because females mate only a limited number of times. Characters that increase a male's rate of obtaining copulations, for example patterns of search behavior and display rate, would be favored by sexual selection in this context. Second, males may attempt to increase their mating success by interfering directly with other males (interference competition). Male-male competition does not usually involve combat in guppies (pers. obs.; Farr 1975). More likely, males may be able to monopolize females at the expense of other males through more subtle interactions such as the jockeying for position described in chapter 2, or by establishing dominance relationships with other males that affect the outcome of interactions. Aquarium experiments suggest ways in which such interactions could affect mating success.

In his competitive mating experiments (see above), Farr (1980b) attempted to tease apart the effects of aggressive behavior and courtship. These experiments involved two-male, two-female groupings. In these and other studies (e.g., Bruce and White 1995), when only two or three males are present, aggressive interactions are commonly observed and one male usually becomes clearly dominant. This may be because male guppies are able to maintain a consistent relationship with one or two other males, in effect remembering the history of past interactions, but are unlikely to form consistent patterns of interactions when more males are present in natural social groupings because they are unable to remember individuals. Viewing interactions between males as a kind of game, it may be that each male's behavior depends on recent interactions with other males. When only one or two other males are present, a male's behavior may become consistently dominant or subordinate. When many males are present, the effects of interactions are much more diffuse, so that dominance relationships are less obvious if they exist at all.

Farr's (1980a) results show that male-male interactions do affect mating success. As noted above (fig. 3.1), the more frequently courting male obtained a disproportionate number of matings in his pairings. However, the mating success of the most frequently courting male was reduced if the other male performed gonopodial thrusts at a higher rate, but not if the other male was more aggressive. In fact, the success of the more frequently displaying male was reduced if he himself was more aggressive. These results suggest that being able to monopolize females is a key factor in this experiment. Males with higher display rates may be better than their competitors at obtaining exclusive courtship interactions with females. This alone could explain their disproportionate mating success, even without female preference for high courting rates. Males are less successful if their courtship is interrupted by the other males' attempts at sneak copulations or if they are frequently involved in aggression toward the other male. Aggression on the part of the more frequently courting male could be as much a sign of attempted interference by the other male as an intrinsic trait of the male himself. Clearly, additional, detailed observations are needed to better understand the role of interference competition between males in sexual selection.

In my competitive breeding experiment (Houde 1988b), I attempted to examine the roles of male-male competition as well as of behavioral preferences of females simultaneously. My design differed from Farr's in that I observed established groups consisting of six males and twelve females. Although I was able to document that the behavioral preferences of females affected the mating success of males, I found no evidence for an effect of male-male interactions on mating success. This is perhaps not surprising,

because interactions between males were uncommon, and clear examples of direct aggression were almost never seen. Brooks (1996b) had similar findings with South African guppies: even though female responses predicted male mating success, male-male competition was uncommon and did not predict mating success.

On the other hand, Kodric-Brown (1992, 1993) was successful in demonstrating that both female choice and male-male competition could influence male mating success. When males were allowed to interact, their mating success was influenced not only by the preferences shown by females in previous dichotomous choice tests, but also by relative dominance, measured by subsequent aggressive interactions. Dominance did not appear to be related to body size, color pattern, or the preferences of females. Dominant individuals obtained more matings than subordinate individuals, even when they were not preferred by females (Kodric-Brown 1993). In thirty-nine "contests" between two males, females preferred the dominant male in twenty-five cases, and the dominant male sired the offspring in almost all of the broods produced (twenty-three of twenty-five). In the remaining fourteen cases, females preferred the subordinate male. In these cases, the dominant (but nonpreferred) male sired all of the sons in five cases, and both males contributed to the brood in nine cases (Kodric-Brown 1992, 1993). Thus, despite the females' preference for the subordinate male, dominant males still showed a mating advantage, although they were not able to monopolize the paternity of the broods. Interestingly, the subordinate males sired a significantly greater fraction (59%) of the offspring in the split broods than did the dominant males (Kodric-Brown 1992). These results suggest that females are able to implement their mating preference through multiple matings and possibly through sperm competition even when they are unable to avoid mating with a dominant male.

The conclusions of these experiments (Farr 1980a; Houde 1988b; Kodric-Brown 1992, 1993) provide only preliminary information about the role of male-male competition in guppies. Most observations of natural social groupings suggest that dominance and aggression are not normally a very important factor, but we need more detailed observations and experiments to determine the effects of subtle interactions between males on mating success.

3.5 Summary

Guppies provide a classic example of a system in which sexual selection has led to the evolution of conspicuous traits that are otherwise detrimental to survival. Conspicuous color patterns and courtship displays of males appear to be favored by female choice, but have the disadvantage of increasing risk of predation. Much of the work on guppies has been devoted to demonstrating that sexual selection occurs and to evaluating the effects of female choice and male-male competition on male mating success, and hence their contribution to sexual selection.

Studies of guppies show that males differ in mating success and that these differences are related to variation in color patterns. There is less clear-cut evidence that differences in mating success are related to male display behavior and other traits. There is substantial evidence for female choice as the mechanism of sexual selection, and especially for mating preferences based on male color patterns. Numerous studies have shown female preferences based on various color pattern elements, although the use of red-orange coloration as a cue has been documented most frequently. There is a good deal of variation among studies, however. Some of this may be due to differences in experimental conditions, but some may reflect genetic differences in patterns of mate choice among guppy populations (see chapter 6). Despite difficulties of interpretation, females also appear to have preferences based on the courtship behavior of males. The differences in female responses to different males, from which we infer mating preferences, have also been shown to affect male mating success, and thus lead to sexual selection on the preferred color pattern characters.

Male-male competition may also affect male mating success, but the specific mechanisms of male-male competition are not well understood in guppies. Scramble competition for access to females, dominance interactions, and direct interference may all contribute to male-male competition in guppies. The roles of specific traits, such as body size, color pattern, display behavior, and aggressive behavior in male-male interactions have yet to be worked out. The fact that there is relatively little overt aggression among males in guppies suggests that the contribution of male-male competition to the overall pattern of sexual selection is probably minor relative to other species.

Given the conclusion that the mating system of guppies is primarily based on courtship by males and mate choice by females, we now turn to the specific mating strategies of males (chapter 4) and the mechanisms of mate choice in females (chapter 5).

4 Male Courtship Behavior

4.1 Courtship Decisions

The seemingly never-ending pursuit and courtship of females by males is one of the most striking characteristics of guppies. Finding and attracting receptive females is one of the key goals in the life history of male guppies. Finding and choosing males to mate with is an analogous problem in the life history of female guppies (see chapter 5). Guppies give us especially good insights into these problems of mate searching, courtship, and mate choice because they show evolutionarily labile and behaviorally plastic responses to variation in the biotic, abiotic, and social environment around them. In this chapter, I provide a general framework for understanding problems of mate finding and courtship faced by guppies in terms of the costs and benefits affecting any life history characteristic, and I examine patterns of variation in the courtship behavior and alternative tactics used by males in light of this framework.

The sequences of behavioral decisions involved in finding, choosing or attracting mates are probably best viewed in terms of trade-offs between current and future costs and benefits in the same way that other life history characteristics have been understood (Stearns 1992). For guppies, each of these activities may affect current reproductive success directly, or may affect future survival or reproductive success through energetic costs and risks of predation.

Behavioral decisions about mate searching and courtship probably have greater consequences for the fitness of males than females, but this needs further empirical documentation in guppies and other species. As in most species with typical sex roles, male guppies have less energetic investment per offspring than do females, the operational sex ratio (sex ratio of receptive individuals) is male biased, and the potential reproductive rate of males is much greater than that of females. All these differences are expected to contribute to greater variance in reproductive success and a greater intensity of sexual selection among males than females (Bateman 1948; Emlen and Oring 1977; Clutton-Brock and Vincent 1991; Clutton-

Brock and Parker 1992). Male guppies are therefore expected to invest heavily in mate-searching and display behaviors that affect their fitness directly though effects on mating success. Female guppies, on the other hand, generally have no difficulty finding a male to fertilize their eggs, so their reproductive success is limited primarily by the rate at which clutches can be produced. The potential for fitness gains for females through mate-searching and mate-choice behavior is probably relatively small and investment in these behaviors may be limited due to the high energetic cost of egg production. It would be interesting to have detailed quantitative data on the relative apportionment of reproductive effort into physiological (i.e., eggs and sperm) versus behavioral components in male and female guppies.

4.2 Patterns of Variation in Male Courtship Behavior

For males, a basic reproductive decision is how much time, energy, and risk to put into courting females. The expected benefit of a male's total courtship effort depends on his rate of encounter with receptive females, his success in getting them to mate with him, the number of ova he fertilizes per female, and, possibly the quality (or viability) of the resulting offspring. Each of these components of reproductive success may be affected in turn by a variety of environmental factors, discussed below.

The costs of a male guppy's courtship efforts depend on how his behavior affects his survival and future reproductive success. All aspects of mate searching and courtship behavior are likely to result in an increased risk of attracting and being eaten by a predator. Displays themselves make males conspicuous to predators as well as to females (Endler 1987). Simply following females puts males at some risk relative to remaining still or hidden. All aspects of mate searching and courtship also require energetic expenditure, both in terms of loss of energy reserves and loss of time spent foraging. The energetic cost of reproductive behavior in male guppies probably acts mainly through direct effects on survival (e.g., risk of dying of starvation), rather than on future reproduction. This is because adult male guppies, unlike females, devote little or none of the energy they obtain from food to enhancing future success through growth (Reznick 1983; Snelson 1989; see discussions of body size in other chapters). In some of the other poeciliids, males show more postmaturational growth than do guppies (Snelson 1989), and body size may have greater effects on mating success (see chapter 3). In these species there may be a trade-off between investment in current courtship behavior and future mating success.

The trade-offs affecting courtship behavior are best explored by observing how particular behavioral elements vary in response to changes in environmental and social circumstances and in the internal state of the individual male guppy. Environmental and social factors may affect the balance between finding and attracting mates and the costs and risks incurred by performing these behaviors. Variation in courtship behavior may be expressed as conditional plasticity in the behavior of individuals (fig. 4.1B), or as genetically hardwired differences among populations (fig. 4.1C).

The total courtship effort of a male guppy is the product of several different elements (fig. 4.1A), each of which is likely to be affected by sexual selection and the potential costs described above. These elements make up the male's overall searching and courtship strategy and may affect mating success in different ways. The most frequently studied elements of courtship include the time and energy spent seeking and following females, the frequency of courtship displays performed to females, the duration and intensity of displays, and actual copulations. In addition to actual displays, the frequency of gonopodial thrusts (sneak copulation attempts) contributes to the overall courting effort. Courtship and gonopodial thrusting may be affected by risks and energetic constraints in similar or in very different ways. These are discussed in detail below. Some behaviors, such as the chases of a single female by two or more described in chapter 2, affect the outcome of interactions among males and are clearly subject to sexual selection through the mechanism of male-male competition. Other behaviors that are features of the courtship display itself are subject to sexual selection through active female choice.

A number of aspects of male mate-searching behavior are more difficult to define in terms of either male-male competition or female choice. For example, the fraction of time a male spends courting should affect his mating success directly but is difficult for females to use as a cue in mate choice, nor is courtship time directly related to interactions between males. Instead, searching for mates is probably better thought of as similar to foraging. Male guppies seeking females are analogous to predators seeking prey in that they seek to maximize the number and quality of females found. In addition to just finding females, males are also interested in finding the most "profitable" females—those most likely to mate. A male's rate of courtship display and of gonopodial thrusting per unit time can both affect mating frequency directly. For example, given that a sigmoid display or a sneak copulation attempt has a certain probability of leading to successful copulation, then an increase in the rate of either behavior would tend to increase mating frequency (subject to various costs and constraints). The rate with which a male displays to an individual female may also affect his attractiveness to her and his likelihood of mating with her (e.g., Farr 1980a). I suspect that overall display frequency has a greater effect on mating success through an increased frequency of mating opportunities (scramble competition among males than through an increased likelihood that the female will copulate following a given display, i.e., female choice).

A. Components of male sexual behavior

Time following female
Frequency of displays
Intensity of displays
Duration of displays
Frequency of copulations
Frequency of gonopodial thrusts

B. Plasticity in male sexual behavior

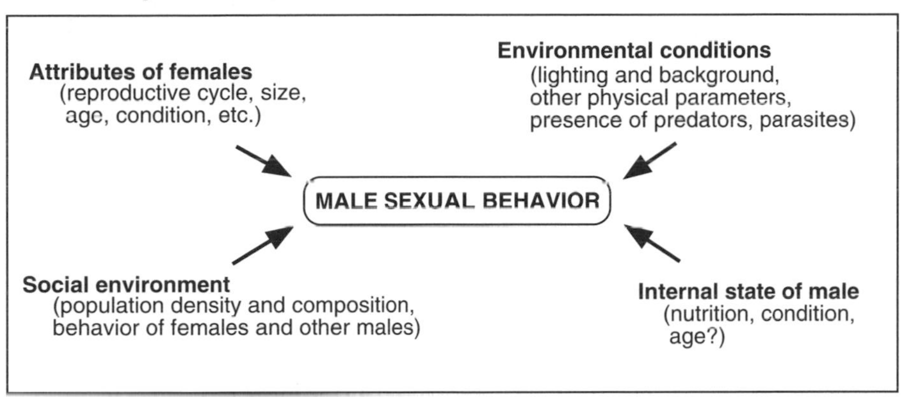

C. Selection on male sexual behavior

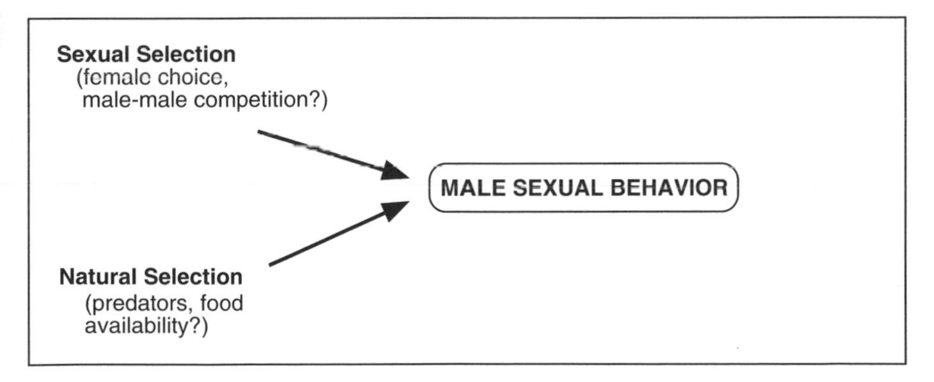

Figure 4.1 Plasticity and selection effects on male sexual behavior. (A) Components of male sexual behavior. (B) Factors leading to plasticity in the expression of male sexual behavior. (C) Factors leading to selection and adaptive divergence in male sexual behavior.

Effect of Energetic Costs on Courtship Behavior

Given that courtship displays are energetically costly, variation in effort devoted to display should depend on the long-term and short-term energetic or nutritional status of males. Energetic considerations might also affect the relative amount of courtship versus gonopodial thrusting performed by males.

Abrahams (1993) modeled the effect of a male guppy's current energetic state on the amount of time spent courting. He predicted that initially hungry males, given a choice between feeding and courting, should feed first until they have obtained enough food to reduce the risk of starvation and then court females. Where food is available at a higher rate, males should be able to increase their energy reserves more quickly and should begin courting sooner. Where the relative benefit of courtship is greater (e.g., with larger females), males should also begin courting sooner.

The experimental data generally supported these predictions. Abrahams gave males in groups a choice between feeding at one end of an aquarium and courting a female at the other end. He observed the sequence of switching between feeding and courting. Males tended to feed before courting, and tended to court sooner when more food was available, as predicted by the model. This variation in courtship time appears to be the result of variation in the risk of starvation. When food levels were low, male guppies had to feed longer or risk starving, but when food levels were high, the males were satiated sooner and could switch to courtship sooner without incurring a high risk of starvation. Also as predicted by the model, relatively more males courted large females than courted small females through the course of the experiment, but the size of females did not affect time spent courting relative to foraging, as would be predicted. The trade-off between courtship time and feeding time thus appears to be more sensitive to differences in the availability of food than to the potential fecundity of the female. This may be because the probability of mating is always so low that the expected difference in payoff from courting a large versus a small female is less than the fitness effect of feeding versus not feeding.

Male guppies in the field also show a trade-off between courtship and feeding (Rodd, pers. comm.). In some, but not all streams sampled, males that spent more time foraging tended to be less likely to invest time and energy in courting the females they encountered. This suggests that the current energetic state of a male may affect his priorities in deciding whether to court a female or to continue feeding.

The idea that the energetic state (or condition) of an individual should affect the effort put into courtship is also supported by work on the effects of parasite infection on the behavior of male guppies. Kennedy et al. (1987) observed courtship by male guppies infected with *Gyrodactylus* (see chap-

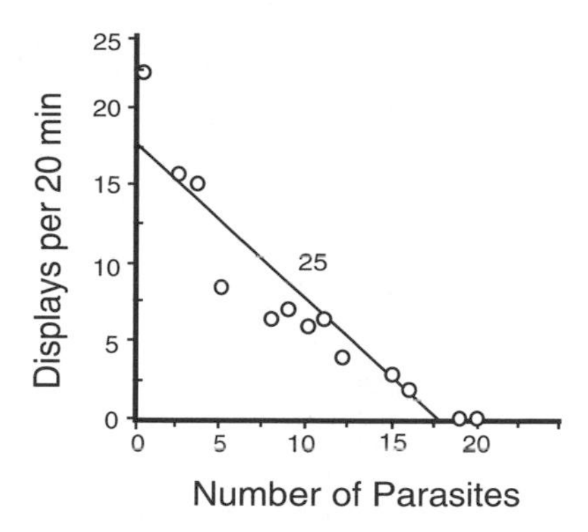

Figure 4.2 Effect of parasite burden (number of parasites per individual guppy) on display rate of male guppies. (Adapted from Kennedy et al. 1987.)

ter 6) and found a strong negative correlation between the number of parasites present and the rate of display (fig 4.2). Interestingly, the maximum number of parasites observed for this analysis was only twenty. In my own work with *Gyrodactylus*, guppies with fewer than twenty parasites generally show no sign of illness. Thus, the results of Kennedy et al. (1987) show that male guppies modify their behavior in response to even the early stages of infection. This suggests that reducing courtship effort may be an attempt to conserve energy stores critical to combating and surviving the parasite infection. In cases of more severe infection or infection with larger parasites such as *Camellanus cotti* (see Kennedy et al. 1987), changes in behavior could be direct consequences of energetic losses to the parasites.

Further evidence that more general aspects of condition affect male courtship behavior come from studies of the effects of toxic substances and other stresses on display rates. For example, courtship is a sensitive indicator of the presence of pollutants (Schröder and Peters 1988), and has also been used to detect heritable effects of mutagens (e.g., Spieser and Schröder 1980; Werner and Schröder 1980). Presumably, the immediate effects of toxic substances and even the inherited effects of mutagens are ultimately mediated in terms of energetics and physiological efficiency and result in reductions in energetic investment in courtship display.

Predation Risk and Courtship Behavior: Behavioral Plasticity

Guppies provide one of the best examples of adaptive adjustment of reproductive behavior in response to predation risk (for other examples, see Tuttle and Ryan 1982; Ryan et al. 1982; Sih 1988; Sih et al. 1990; Travers and Sih 1991; Forsgren and Magnhagen 1993; Chivers et al. 1995). Guppies adjust their courtship behavior through behavioral plasticity and through adaptive genetic variation among populations. Studies have examined the

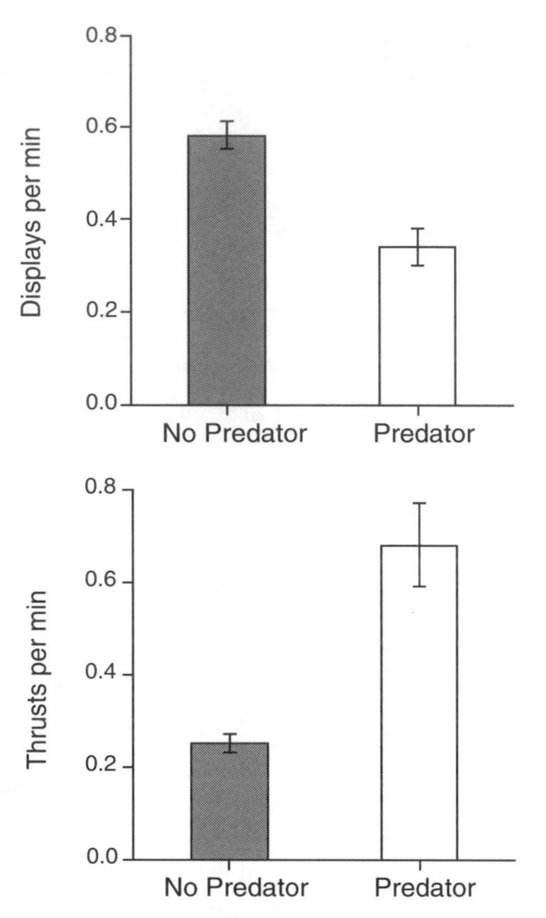

Figure 4.3 Effect of the presence of a predator (*Crenicichla*) on the rate of sigmoid display and gonopodial thrusting (sneak copulation attempts). Error bars indicate one standard error. (From data in Endler 1987.)

effect of the actual presence of predators on courtship (Magurran and Seghers 1990c; Magurran and Nowak 1991; Godin 1995), the effect of conditions that increase the risk of predation on courtship (Endler 1986; Reynolds 1993; Reynolds et al. 1993), and the effect of variation in the risk of predation among populations (Ballin 1973; Farr 1975; Crow 1981; Luyten and Liley 1985; Magurran and Seghers 1990c, 1994c; Rodd and Sokolowski 1995).

The most direct demonstrations that risk of predation affects courtship behavior come from observations on the effect of the actual presence of a predator on courtship behavior, and the results are unambiguous. Endler (1987) compared courtship behavior of male guppies over several months

(November–March 1982–1983) when no predator in artificial greenhouse streams was present to courtship in one month (April 1983) when a predator (the cichlid *Crenicichla alta*) was present. Rates of sigmoid display declined between the predator-free and predator-present time periods, while rates of gonopodial thrusting increased (fig. 4.3).

Similar kinds of changes have been demonstrated by Magurran and coworkers using more strictly controlled observations. In two separate experiments (Magurran and Seghers 1990c; Magurran and Nowak 1991), the presence of a predator resulted in a reduction in the frequency of sigmoid displays and an increase in the frequency of gonopodial thrusting. Magurran and Nowak (1991) suggest that males increase the use of gonopodial thrusting as a way of exploiting females' preoccupation with observing and avoiding the predator. The changes in courtship behavior of males occurred in lab-reared stocks of a high-predation population from the Oropuche River in response to *Aequidens pulcher*, a cichlid (Magurran and Nowak 1991), and in wild-caught males from a high-predation locality in the Lower Aripo River, in response to *Astyanax bimaculatus*, a characin (Magurran and Seghers 1990c). There was no such response in guppies from a low-predation locality in the Aripo River (Magurran and Seghers 1990c), indicating genetic differences in the degree of behavioral plasticity among populations.

The results of these laboratory experiments were corroborated by an experiment conducted in the field. Godin (1995) presented a model predator (*Crenicichla alta*) suspended from a rod to groups of guppies in a locality in the Quare River of Trinidad where *C. alta* are common. As in the previous studies, males reduced their rate of sigmoid courtship displays but increased the rate of gonopodial thrusting when the model predator was present. Godin (1995) suggests that the change in male behavior might have been a direct adaptive response to the perceived risk of predation, but might also have been a response to changes in the behavior of females in response to the model predator. Sneak copulation attempts may be more successful when females are preoccupied with inspecting a predator, or sigmoid displays may be less successful if females reduce their responsiveness to males when a predator is present. Further experiments (Godin, pers. comm.) revealed, startlingly, that the most colorful males reduce their display rates in response to a predator relatively more than the less-colorful males. This result suggests that colorful males "know" that their risk of predation is high relative to less colorful males, but begs the question of how they "know" this.

Guppies also appear to modify their courtship behavior in response to variation in light level, apparently because light level is positively correlated with predation risk. Endler (1987) found higher rates of sigmoid dis-

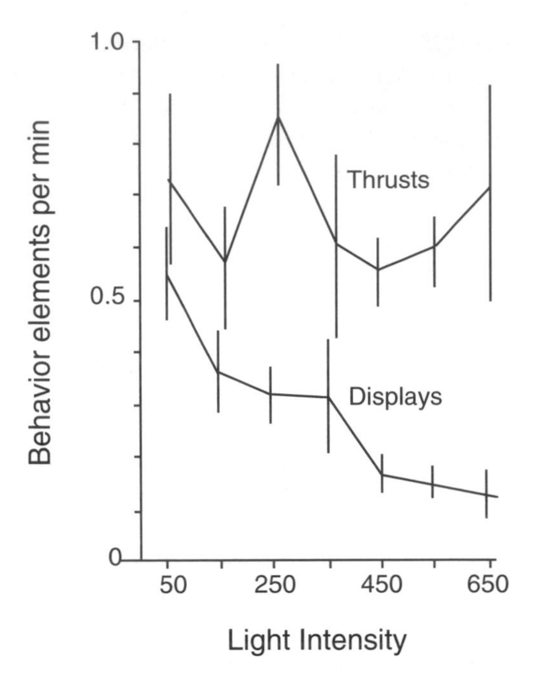

Figure 4.4 Effect of light level on rates of sigmoid display and gonopodial thrusting (sneak copulation attempts). Error bars indicate one standard error. (Adapted from Endler 1987.)

plays in the winter observation period with predators present (April 1983) than in the subsequent summer period (May–October 1983), also with predators present (fig. 4.4). He attributed this decline to a seasonal increase in light levels. There were also pronounced diurnal changes in courtship rates in the summer, with decreasing rates of sigmoid displays at high light intensities at midday. The predator *C. alta* was also most likely to attack guppies at high light levels. Light levels did not appear to affect rates of gonopodial thrusting, however. The changes in courting rate suggest that the guppies may perceive a high risk of being preyed upon at high light levels and adjust their display behavior accordingly.

Reynolds and coworkers took this idea one step further and argued that they could manipulate guppies' perceived risk of being preyed upon by experimentally manipulating light levels (Reynolds 1993; Reynolds et al. 1993). The result of manipulating light level was a reduction in rate of sigmoid display (fig. 4.5), similar to the findings of Endler (1987). The perceived difference in predation risk appears to be shaping the courtship behavior of male guppies in this experimental system. However, only large males reduced their rate of display under high light conditions. A previous study (Reynolds and Gross 1992) had shown that female guppies from the same study population had a preference for larger males, so both male size

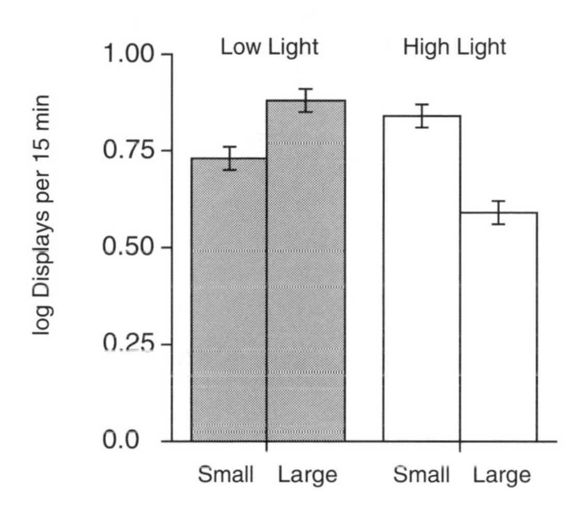

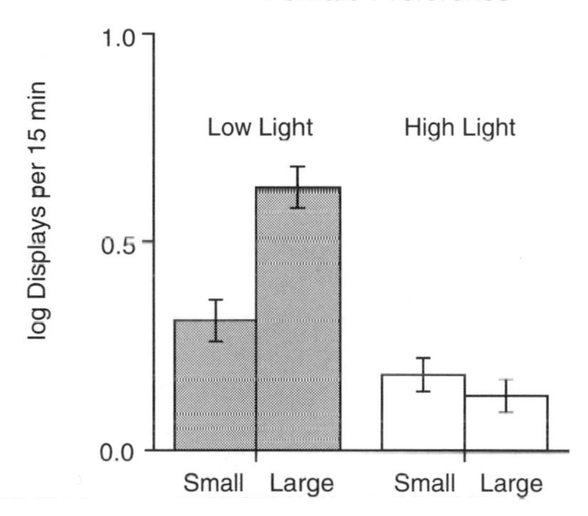

Figure 4.5 Effect of male size and light level on rate of sigmoid displays and female responses. Error bars indicate one standard error. (Adapted from Reynolds et al. 1993.)

and courtship behavior could be subject to natural and sexual selection. Under low light, large males displayed more frequently than did small males, but the reverse was true under high light.

There are two possible explanations for this interaction between light level and male size. Reynolds (1993) argued that increasing the light level should increase the relative conspicuousness of large males more than that of small males. This change in conspicuousness could affect the balance

Table 4.1
Summary of Studies on Effects of Predation on Courtship

Reference	Location	Relative Courtship Rate (High Predation vs. Low Predation)
Farr 1975	Field	Hi pred > lo pred
Luyten and Liley 1985	Field and lab	Hi pred < lo pred
Magurran and Seghers 1990c	Lab without predator	Hi pred > lo pred
	Lab with predator	Hi pred = lo pred
Magurran and Seghers 1994c	Field	Hi pred > lo pred
Shaw et. al. 1994	Field	Hi pred = lo pred
Rodd and Sokolowski 1995	Lab	Complex effects
Houde and Cassidy (unpubl. data)	Lab	Hi pred < lo pred*

NOTE: Except as indicated, *Crenicichla* is the major predator in high predation localities, *Rivulus* is the major predator in low predation localities.
*High predation = *Agonostomus/Eleotris* (Yarra River); Low predation = *Rivulus* (Paria River).

between costs and benefits of courtship in two ways: the relative risk of courting increases more for large males as the light level increases than it does for small males ("differential cost"), and the relative attractiveness of large males to females increases more as light level increases than it does for small males ("differential benefit"). In both cases, large males are expected to reduce their amount of courtship relative to small males under high light. The differential cost hypothesis suggests that large males should reduce their courtship at high light levels because courtship becomes relatively more costly as a result of increased conspicuousness to predators, and predicts that large males' mating success should decrease relative to that of small males at high light levels. The differential benefit hypothesis suggests that large males should court less at high light levels because they become relatively more attractive to females, and predicts that large males' mating success should increase relative to that of small males at high light levels. These hypotheses are not necessarily mutually exclusive, and could have simultaneous and opposite effects on mating success.

In fact, Reynolds (1993) found a decrease in mating success of large males relative to small males at high light levels, consistent with the prediction of the "differential cost" hypothesis. This result does not exclude the possibility that increased conspicuousness increases the relative attractiveness of large males under high light as argued under the "differential benefit" hypothesis, and this idea deserves to be explored further. The experiment suggests that the change in predation risk has the more important effect, however.

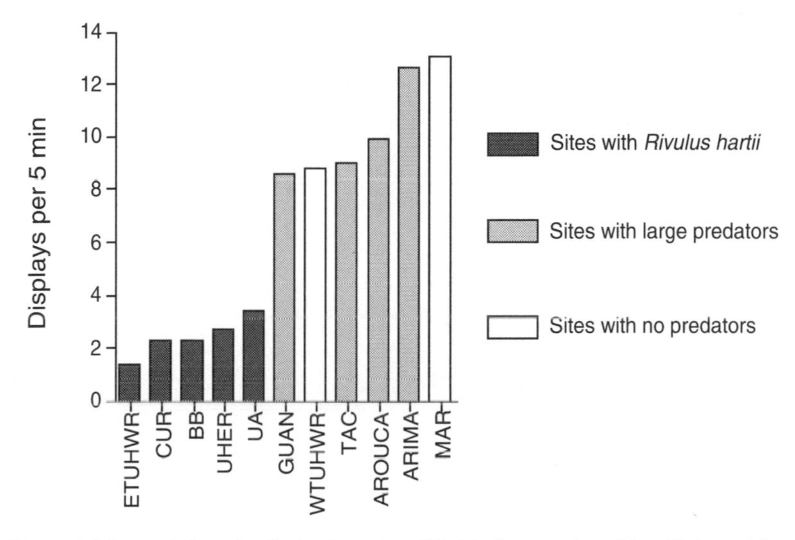

Figure 4.6 Rate of sigmoid display in various Trinidad stream localities. (Adapted from Farr 1975.)

Predation Risk and Courtship Behavior: Genetic Variation among Populations

There is also evidence for differences in courtship behavior between populations that experience different regimes of predation, but results are incon sistent among studies (table 4.1). In a field study, Farr (1975) found significant differences in display rates among populations of male guppies (fig. 4.6), which he attributed to differences in both predation regime and social structure (see more on social effects on courtship below). Farr (1975) found that in localities with large predators (*Crenicichla*) and in most localities with no predators, male guppies showed high rates of courtship. On the other hand, in localities with *Rivulus*, male guppies showed low rates of courtship even though *Rivulus* is generally thought not to be a dangerous predator (Haskins et al. 1961; Endler 1978; but see Seghers 1973). Farr suggested that the sit-and-wait hunting behavior of *Rivulus* could make displaying especially dangerous for male guppies in *Rivulus* localities. He also pointed out that in all but one of the *Rivulus* localities, guppies lacked social cohesiveness and that this, rather than the presence of predators, could have contributed to low courting rates, although the expected relationship between group cohesiveness and courting rate is not clear.

Magurran and Seghers (1990c) observed similar differences among guppies from *Crenicichla* and *Rivulus* localities in a laboratory situation with no predator present, but no difference in display rate when a predator was present (see discussion of plasticity in courtship above). The laboratory

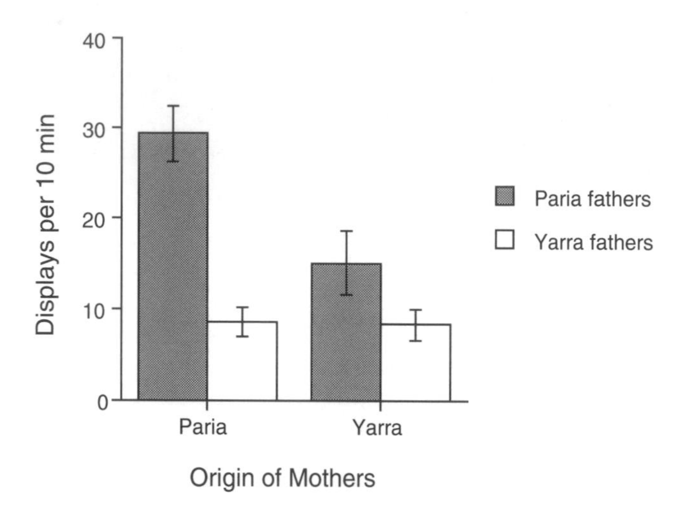

Figure 4.7 Rate of sigmoid display in lab-reared males from Paria and Yarra populations and hybrids from these populations. Error bars indicate one standard error. (Houde and Cassidy, unpublished data.)

experiment controlled for effects of density and sex ratio that were uncontrolled in Farr's observations. Subsequent field observations (Magurran and Seghers 1994c) bear out the pattern of higher combined rates of display and thrusting in *Crenicichla* localities. Shaw et al. (1994), on the other hand, found no significant differences in the display and thrusting rates of males in a field study of two other populations. Unlike Farr (1975), Magurran and Seghers (1994c) suggest that the differences may be the result of several ecological factors. Male guppies spend relatively less time feeding in high predation sites and more time following females. This may be due to greater productivity and food availability in high predation sites or lower energetic requirements of those males because of smaller body size. High rates of courtship could also reflect high, early reproductive effort in life histories evolved under risk of predation.

Luyten and Liley (1985) also demonstrated population differences in the rate and duration of sigmoid displays in the field, and showed that these behavioral differences persisted in laboratory-reared offspring of wild-caught fish. The differences that Luyten and Liley (1985) found were at odds with Farr's (1975) and Magurran and Seghers's (1990c, 1994c) observations, however. Observations by Luyten and Liley (1985) and by previous students in Liley's group (Ballin 1973; Crow 1981) all indicate that male guppies display at higher rates in *Rivulus* localities than in *Crenicichla* localities, while Farr (1975) and Magurran and Seghers (1990c, 1994c) found the opposite.

Finally, a comparison of two guppy populations from the "Caribbean" predator fauna (Endler 1983; see table 1.1) carried out in my laboratory revealed that male guppies from a high-predation population (Yarra River) displayed at a significantly lower rate than did males from a low-predation population (Paria River) (fig. 4.7; Houde and Cassidy, unpublished data). This result is most consistent with Luyten and Liley's (1985) study.

The reason for the discrepancies between results of the various studies remains unclear, although the large number of factors that seem to affect male courtship behavior makes variation between studies almost inevitable. While there appear to be genetically based differences in courtship behavior indicated by results with lab-reared fish, rearing conditions and observation conditions may have a large influence on the outcome of studies.

Part of the reason for variation in results between studies is that genetic differences and conditional responses in male courtship behavior interact in complicated ways (i.e., gene-by-environment interactions). Magurran and Seghers (1990c) and Rodd and Sokolowski (1995) have examined genetic and environmental effects on courtship simultaneously. Magurran and Seghers (1990c, see above) found that response of male guppies to the presence of a predator differed between high- and low-predation localities. Those from the high-predation locality reduced their rate of sigmoid displays and increased their rate of gonopodial thrusting when a predator (*Astyanax*) was present, but those from the low-predation locality where *Astyanax* does not occur did not alter their courtship behavior when the predator was present (fig 4.8).

Rodd and Sokolowski (1995) looked at the relationship between body size and courting rate in males from stocks originating from high- and low-predation localities. In a striking parallel with the results of Reynolds et al. (1993), they found a positive relationship between body size and courting rate in males from the low-predation stocks but a negative relationship in males from the high-predation stocks, but no overall effect of origin on courting rate. A similar explanation may apply to Rodd and Sokolowski's (1995) results as to those of Reynolds et al. (1993). In populations that experience a high risk of predation (*Crenicichla* stocks in Rodd and Sokolowski's study) or when the risk of predation is perceived to be high (Reynolds et al.'s high light treatment), large male guppies are at greater risk of predation than are small males and are expected to reduce their rate of display. Rodd and Sokolowski (1995) also found complex interactions between genetic effects of population of origin and the social environment in which fish were reared (see below). Both Rodd and Sokolowski's (1995) work and the study by Magurran and Seghers (1990c) demonstrate that variation among populations is not necessarily a matter of simple differences in mean values of characters. Instead, phenotypically plastic behav-

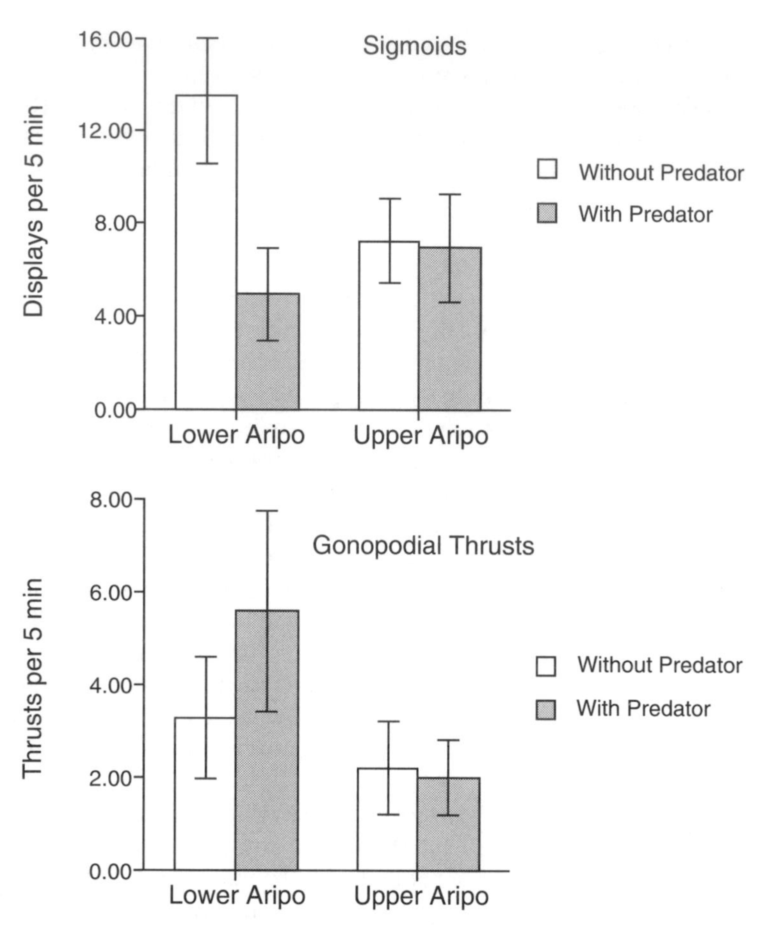

Figure 4.8 Display rate of high-predation (Lower Aripo) and low-predation (Upper Aripo) male guppies in the presence or absence of a predator (*Astyanax*). Error bars indicate 95% confidence intervals. (Adapted from Magurran and Seghers 1990c.)

ioral characters such as courting rate may show more complex gene-by-environment interactions.

Genetics of Courtship Behavior

While comparisons of courtship behavior in lab-reared guppies provide evidence that population differences have a genetic basis, there is relatively little information about the genetic details of these differences. In a preliminary analysis of hybrids between the Paria and Yarra populations carried out in my laboratory (Houde and Cassidy, unpublished data), all hybrids between the two populations displayed at a low rate, similar to that of pure

Yarra males (see fig. 4.7). This suggests a dominant genetic effect. There is a suggestion, however, that hybrids with Paria fathers and Yarra mothers may have a somewhat higher display rate than the reciprocal hybrid, possibly a result of maternal- and/or sex-linked effects. More detailed analyses of F_2 hybrids would be necessary to determine if this is the case.

In addition to genetic differences in male courtship behavior among populations, there is almost certainly considerable genetic variation in this and other behavioral characters within populations. Farr (1980b) documented differences in courtship behavior between the inbred strains he used in his behavioral studies. In a series of crosses between strains, he then showed that almost all the differences in courtship had some Y-linked component, as do the associated color pattern differences (Farr 1983, 1984). The question remains whether the kinds of differences Farr (1980b, 1983, 1984) documented between strains are representative of individual variation in natural populations. Conceivably, random fixation of alleles in the inbred strains during their history of domestication in the laboratory could have contributed to the differences in behavior. Quantitative genetic analyses of variation in courtship behavior of natural populations are warranted.

Social Interactions and Courtship

The presence and behavior of conspecifics also affects male courtship behavior. In other species, mature males appear to suppress the maturation of younger males, effectively their courtship displays (Borowsky 1973; Snelson 1989; Bushmann and Burns 1994). A pilot study suggests that the opposite may occur in guppies: males matured more quickly when they were in view of other adult individuals (Houde, unpublished data). Male guppies tend to increase their display rate when other males are present (Farr and Herrnkind 1974; Farr 1976, 1980b), although the greatest difference seems to be between the display rates of single males versus more than one male (Farr and Herrnkind 1974). The presence of other males affects the courtship behavior in other poeciliids, although the effect is often size dependent (Travis and Woodward 1989; Zimmerer and Kallman 1989). These changes in male behavior could be in direct response to the presence of the competing male, or could reflect changes in the way the female behaved in interactions with both males. Another possibility is that social effects on courtship are mediated by hormonal changes in males (e.g., Groothius 1992).

Characteristics of potential competitors may affect how males modify their courtship as well. In Farr's (1980b) study, male guppies appeared to increase their display rate when another male was present to a level that matched that of their competitor. Darwall (1989) found that the rate of a male's display decreased with increasing absolute size of the competitor. In addition, the frequency of display increased as a male's size relative to the competitor increased. In mollies, the effect of potential competitors on courtship rates were independent of the relative sizes of the males (Travis and Woodward 1989).

Besides the direct effects of competing males on courtship behavior, there may be more subtle effects of the social environment in which male guppies were raised. Rodd and Sokolowski (1995) examined the effects of the demography (density of females, males, and juveniles) and origin (*Rivulus* or *Crenicichla* locality) of the population in which males were raised on their courtship behavior. The demography of a male's home population could affect the level of competition and opportunities for courtship he has experienced. Experimental populations formed from *Rivulus*-locality or *Crenicichla*-locality fish may have had different behavioral interactions with the focal males. Rodd and Sokolowski's (1995) results showed an effect of both demography and origin, although the exact ways in which the courtship behavior of the focal males varied was not clear-cut or easily explained owing to complex interactions between the variables. The general implication, however, is that male guppies are able to fine-tune their courtship behavior in response to the availability of mating opportunities or to the severity of competition with other males. This is supported at least in fish derived from *Rivulus* localities which increased their courtship effort in response to a high density of competing males and reduced their courtship effort in response to a high density of available females.

An even more subtle way that male guppies adjust their courtship to their social environment has recently been documented by Dugatkin and Sargent (1994). They allowed males to observe other males, either in proximity to a female or not, and showed that these males later preferred to associate with males that had not been in proximity to a female. Dugatkin and Sargent suggest that in this way, males are able to avoid being near other males that are particularly attractive, and which would reduce their own chances of attracting a female. An alternative explanation could be that males are avoiding being near other males that are good competitors and able to monopolize females, also reducing chances of successful mating.

All of these studies show that male guppies are very sensitive to their social environment and are able to adjust their courtship in apparently adaptive ways.

4.3 Male Mate Choice: Effects of Female Characteristics on Male Courtship Behavior

A final source of variation in the display frequency of male guppies is the behavior, reproductive status, and size of the females with whom they interact. These attributes of females affect the relative costs and benefits of sexual behavior for males, leading to variation in the effort that males put into following and courting different females. This kind of variation in effort by males is termed *male mate choice*.

Given that there is some limitation on their ability to copulate repeatedly, males are expected to exercise some selectivity in their choice of females, especially when females can vary greatly in their quality as mates (Owens and Thompson 1994; Deutsch and Reynolds 1995). Receptive females have a greater probability of mating than unreceptive females, so it is clearly advantageous for males to concentrate their courtship effort on receptive females. In addition, it should be advantageous for males to court larger females because of their greater potential fecundity (e.g., Abrahams 1993), and perhaps because larger females (and their unborn offspring) have a lower risk of being preyed upon.

Even though it does not usually lead to sexual selection on female traits (Fitzpatrick et al. 1995), male mate choice has been the focus of an increasing number of studies (e.g., Shelly and Bailey 1992; Hill 1993; Simmons et al. 1994; Krupa 1995; McLennan 1995; Riechert and Singer 1995; Verrell 1994; Wiernasz 1995; see Andersson 1994 and Johnstone 1995 for more examples). Preferences of males for large females is especially well documented in fishes (e.g., Downhower and Brown 1980; Rowland 1982; Sargent et al. 1986; Côté and Hunte 1989), as are preferences for receptive females (e.g., Hughes 1985; Sumner et al. 1994). In general, males are predicted to be most choosy when operational sex ratios are female biased, i.e., when females are most readily available (Hubbell and Johnson 1987; Shelley and Bailey 1992; Krupa 1995). This appears to be the case in some populations of guppies (Rodd and Sokolowski 1995; F. H. Rodd, pers. comm.). We could also predict that the degree of choosiness might depend, as it does in females (see chapter 5), on predation risk and other costs of engaging in sexual behavior, but this remains to be investigated.

In guppies, Liley (1966) noted a generally greater level of sexual activity of males interacting with receptive females relative to males interacting with unreceptive females. In Liley's study, receptive females had recently given birth to young. Males also seem to prefer newly introduced females over resident females (Heinrich and Schroder 1986). Farr (1980a) compared males that differed in their previous experience with receptive versus nonreceptive females and found no effect on their display rates. This suggests that although males may modify their courtship from moment to moment in response to the behavior of females, variation in female behavior does not necessarily have a long-term effect on male courtship.

The immediate behavior of females does seem to affect the behavior of males, however. Kennedy (1979) examined the effect of isolation on males and females and found that females who had been isolated received more courtship than other females, and that the isolated females tended to approach and remain near males. The isolated females may have been perceived as novel by males, and therefore more likely to be receptive than other females. More likely, however, the isolated females were more receptive to male courtship for psychological and hormonal reasons and could have been actively inciting males to display. On the other hand, once a male has located a receptive female he may actually display less frequently and may be more persistent in following her (Houde 1988b), perhaps reducing the risk of interruption by other males. In a comparison of male courtship directed to receptive and unreceptive females, Hankes and Houde (unpublished data) also found that males had a slightly (but nonsignificantly) lower display rate, but a significantly greater rate of switching between females when males were courting unreceptive females. In general, males appear to pursue receptive females more persistently but, if anything, with a lower total rate of display than they do unreceptive females.

Based on changes in color patterns within individual males, Baerends et al. (1955) argued that males reached a higher motivational state when courting large females, although the association of particular color pattern changes with motivational state described by these authors does not apply to all males. Abrahams (1993) also found that males were more likely to court females than to feed if the females were large (see discussion above), but he argued that food availability probably has a greater effect on courtship than the size of females. In some of my own experiments (e.g., Houde 1987; see chapter 3), males seemed to prefer to court large, unreceptive females over small receptive females even though the small receptive females seemed to solicit courtship from the males. This preference for large but unreceptive females over small receptive females does not seem to make a lot of sense given that unreceptive females are unlikely to mate, but it may be advantageous for males to maximize their encounter rate with large females, a few of whom will turn out to be receptive.

A more clear-cut effect of female size on male courtship appeared in laboratory experiments in which males were given a choice of large and small females (Houde, Hankes, and Stanley, unpublished data). In one such experiment, five males were allowed to court eight females, of which four were large and four were small. In this situation, males directed more displays to large than to small females. However, in another experiment in which the total number of females equaled the total number of males, there was no significant difference in courtship directed toward large compared to small females, probably because once all the large females were being monopolized, other males had no choice but to court small females.

Male guppies also show preferences depending on the genetic background of females and on their own genetic background. Males distinguish among females from mutant strains that differ visibly in background color, tending to prefer females from their own strain (Haskins and Haskins 1950; see also Heinrich and Schröder 1986). This kind of discrimination by males contributes to reproductive isolation between guppies and sympatric congeners (Haskins and Haskins 1949; Liley 1966) and perhaps also to assortative mating among different populations (Ballin 1973; Luyten and Liley 1991). This deserves further study in order to learn more about the origins of reproductive isolating mechanisms (see discussion of speciation in chapter 6).

Finally, females can sometimes be much larger than males and sometimes behave aggressively toward them, potentially reducing male courtship behavior, at least in laboratory situations (pers. obs.). It is not clear if this is a common occurrence in natural populations, though. I have most often observed aggression by females toward males when a large female is housed with a single male for an extended period. Under these conditions the female often chases and nips at the male. The result is that when the male is used in an experiment he often appears clearly fearful even of receptive females and avoids rather than courts them. Any approach by the female, even if sexually motivated, results in a rapid retreat by the male. These males tend to display infrequently and at very low intensity, even after they have been separated from the aggressive female. These changes in the behavior of males may be another example of plastic responses to a changeable social environment, and may even be adaptive, or they may reflect a pathological response to an unnatural social situation. In either case, aggression by females toward males can cause difficulties in interpreting experimental results.

4.4 Sneak Copulation as an Alternative Mating Tactic

Much of the variation in the display rate of male guppies that results from genetic variation and behavioral plasticity is accompanied by variation in the rate of gonopodial thrusting. Gonopodial thrusting, or sneak copulation behavior, can be viewed as an alternative mating strategy of males that is similar to alternative mating tactics or sneaking strategies in other species (Reynolds et al. 1993; see also Austad 1984, Dominey 1984, and Andersson 1994 for further discussion). Sneak copulations avoid some of the energetic cost of displaying but have a relatively low probability of success (Clark and Aronson 1951; Liley 1966; Farr 1980a; Luyten and Liley 1991). Thrusting or sneak copulation attempts are associated with situations in which displays are not expected to be effective or are relatively costly. The effectiveness of displays relative to their costs may depend on the receptivity and mating preferences of females, the presence of competitors, visibility of displays to females, the risk of predation, and other factors.

Unlike guppies, many of the other species in which males use alternative mating tactics show much stronger differentiation among behavioral phenotypes and a strong association between male size and behavior (e.g., Constantz 1975; Borgia 1980, 1982; Thornhill 1979, 1980, 1981; Gross and Charnov 1980; Dominey 1980, 1984; Gross 1982, 1985, 1991a,b).

Several poeciliid provide especially good examples of size-related variation in behavior, especially because the polymorphisms are known to have a strong genetic basis (Constantz 1975; Hughes 1985; Woodhead and Armstrong 1985; Farr et al. 1986; Ryan and Causey 1989; Travis and Woodward 1989; Zimmerer and Kallman 1989; Travis 1994). In poeciliids, at least, the most likely mechanism for the maintenance of the correlated variation in size and behavior is the female preference for large size in males (Hughes 1985; Ryan and Wagner 1987; Bisazza and Marin 1991; Reynolds and Gross 1992; Rodd and Sokolowski 1995). A preference for large males makes advertisement displays profitable for large but not small males; hence sneak copulation is the more profitable tactic for small males. In contrast with species in which size and behavior are polymorphic, individual male guppies use both courtship and sneak copulation, depending on the situation. There is also less evidence for consistent behavioral differences among male guppies relative to other species, although what variation there is has been documented because of its association with body size (Rodd and Sokolowski 1995). Sticklebacks appear to be similar to guppies in that sneaking is a facultative, opportunistic behavior rather than a specialized behavioral phenotype (Goldschmidt et al. 1992).

Male guppies tend to attempt sneak copulations more often relative to the frequency of displays when females are unreceptive (Farr 1980b) and when displays are unlikely to increase a male's chance of obtaining a copulation. Males also seem to thrust rather than display when females are distracted by the presence of predators and are unlikely to notice or respond to displays or to flee from the sneak copulation attempt (Magurran and Nowak 1991; Magurran and Seghers 1994c; pers. obs.). Magurran and Nowak (1991) provide a game-theoretic model predicting that as predation risk increases, females should decrease the degree to which they avoid sneak copulations and perform antipredator behavior instead. As a consequence of the females' decreasing attentiveness to the males, males are predicted to simultaneously increase their relative rate of gonopodial thrusting. Above some level of predation risk, both males and females are predicted to cease all mating activity and to switch to antipredator behavior.

Males also appear to increase their rate of thrusting when displays are especially risky, possibly independent of the behavior of females. The presence of a predator leads to an increase in the rate of thrusting relative to displaying, at least in males derived from high-predation populations (Endler 1987; Magurran and Seghers 1990c; Magurran and Nowak 1991). Males also use thrusting relatively more frequently under high light levels, when displays are most conspicuous to predators (Endler 1987), and in turbid water conditions with poor visibility of males to females (Luyten and Liley 1985).

The rate of thrusting may also depend on the density of competing males. Both Farr (1976) and Rodd and Sokolowski (1995) found that rates of thrusting increased with increasing density of males. The relative sizes of a male and competitor male also affect the rate of thrusting as well as the rate of display (Darwall 1989). Males tend to thrust more frequently when they are small relative to their competitor or when the competitor is large, regardless of the size of the focal male. It is not clear whether being smaller than the competitor makes a male's displays riskier or merely less effective, leading to an increase in thrusting.

All of these examples are consistent with an adaptive pattern of adjustment of relative effort put into displays versus gonopodial thrusting. There is similar variation in the relative emphasis on display versus gonopodial thrusting among poeciliid species (Farr 1989; Bisazza 1993). Some species have no display at all, and always inseminate females by thrusting, while others, like guppies, use display as the most frequent mating strategy. The degree of development of courtship display, color patterns, sexual dimorphism, and female mate choice in guppies can be considered extreme compared to most other poeciliid species. This interspecific variation may be related to costs and benefits of the two strategies similar to those within species. The variation in dimorphism and use of courtship display among species could also reflect ecological costs such as selective predation on conspicuous displays that constrain the evolution of these characters. Guppies, with their conspicuous color patterns and courtship displays, would seem to be relatively free of such constraints compared to other less conspicuous species, but the reasons for the differences are still unclear. Nevertheless, variation of color patterns with predation regime in guppy populations (Endler 1978, 1980, 1983) indicates that there is at least some constraint imposed on the elaboration of color patterns and displays.

One further intriguing parallel is the finding of Reynolds et al. (1993) that rate of thrusting in guppies is positively correlated with gonopodium length, at least under low light conditions. A similar pattern occurs in the Gila topminnow, in which small males engage primarily in sneak copulations and have relatively longer gonopodia than large, courting males (Constantz 1975). These intraspecific patterns mirror an interspecific association between gonopodium length and relative importance of thrusting among poeciliids (Rosen and Tucker 1961). The implication is that a long gonopodium is advantageous for males that depend on sneak copulations, possibly because it improves success in inseminating evasive females.

4.5 Summary

Taken together, studies of the sexual behavior of male guppies provide an excellent example of how specific costs and benefits can affect the expression and evolution of behavior. We assume that the overall pattern of male sexual behavior is under selection to maximize the number of successful inseminations over the male guppy's lifetime. The incessant attentions male guppies pay to females reflect this. In comparison with other species, male guppies devote a great deal of effort to sexual behavior and pursuit of females. However, patterns of variation in the behavior of males indicate that their reproductive strategy has evolved subject to constraints imposed by energetics, time limitations, risk of predation, and the reproductive biology of females. The effects of these constraints can be seen in a number of elements of male courtship behavior in guppies and also in the relative use of the alternative tactics of display and sneak copulation.

Courtship appears to involve significant energetic investment, although this has not been measured directly in any study. Males vary their courtship behavior depending on their current nutritional or energetic state. As a result, there appear to be trade-offs in the time males devote to feeding versus courting and pursuing females. The simplest interpretation is that these trade-offs depend on the male's assessment of the risk of starving in the near future. It is also possible that current courtship efforts affect the long-term energetic state of males in ways that affect future reproduction. Energetic studies of male courtship would thus be of interest.

The interplay between predation risk and courtship behavior is the best-studied aspect of the sexual behavior of males. Obviously, being preyed upon has an immediate effect on future reproduction. Male guppies appear to adjust their courtship behavior in relation to predation risk both facultatively and through evolutionary changes in populations. The presence of a predator in experimental situations generally leads to a reduction in the frequency of courtship displays and an increase in the rate of sneak copulation attempts. Simulating a risky environment by increasing light levels has a similar effect on male courtship. We might expect that differences in predation risk among populations would lead to the evolution of differences in courtship similar to the facultative changes seen in short-term experiments. The results of field and laboratory comparisons of courtship behavior differ among studies, however. Some studies show an increase in courtship in high-predation localities, but others show a decrease. The effects of additional factors such as social structure in field and laboratory populations need to be teased apart in future studies in order to resolve these inconsistencies. It is clear, though, that there are genetic differences among populations in courtship behavior.

A number of social factors affect the courtship of male guppies in complex ways, and these may contribute to the variation in results among studies. The presence of other males, the sex ratio and the age and size structure of the group, and population-specific behavior patterns of males and females all affect the courtship behavior in male guppies. More work needs to be done on the effects of social factors like these, and, in particular, specific predictions need to be made about how social structure affects the availability of mating opportunities, and in turn affects the behavior of males.

It is clear that the effort males can devote to pursuing and courting females is limited and that male guppies therefore show some selectivity in their choice of females. In particular, males prefer large (highly fecund) females and receptive females that respond to their displays. Male mate choice also appears to be implicated in species recognition in places where other *Poecilia* species occur sympatrically. There is some indication that males discriminate among females that differ genetically within *Poecilia reticulata*, but this warrants further study.

Much of the variation in display behavior of male guppies is accompanied by variation in rates of sneak copulation attempts. While displays are energetically costly and risky, sneak copulations are cheaper and safer but less likely to be successful. Thus much of the variation in sexual behavior in guppies can be interpreted in the context of alternative mating tactics that have been studied extensively in other species. Guppies differ from most of these species in that both tactics are part of the behavioral repertoire of most if not all males and are used depending on the immediate situation. It would be interesting to have more information on whether males differ consistently in the degree to which they use courtship versus gonopodial thrusting.

5 Evolution of Female Choice 1: Direct Selection, Adaptive Plasticity, and Sensory Drive

5.1 Mate Choice Decisions

Like males, female guppies must also find and choose mates, but for females, deciding which males to mate with is probably a more important problem than locating potential mates. In chapter 3, we saw that mate choice in guppies appears to affect the evolution of male color patterns and other traits. Female guppies behave as though they prefer to mate with males with attributes such as large amounts of orange in their color patterns, and we can infer that these preferences are likely to have consequences for the evolution of male color patterns. But how, in a proximate sense, does mate choice come about and how is it shaped by selection?

Mate choice is a decision-making process involving sensory and cognitive filtering of information about each male combined with other decisions that depend on a variety of internal and external factors. The behavioral components of mate choice (fig. 5.1A) include the receptivity and responsiveness of females, their degree of choosiness, and their specific preference functions (see below). In this chapter and the next we will view the overall pattern of mate choice and its components as characters which, like the courtship behavior of males, can show flexibility in response to various influences (fig. 5.1B) and can evolve in response to selection (fig. 5.1C). The expression of female mating preferences and the ways in which they are affected by selection depend on how male color patterns are perceived by females and also by predators. Guppies have provided some of the best information available about the interplay between sensory systems and sexual selection.

5.2 Behavioral Components of Mate Choice

The elements that contribute to the overall pattern of mate choice are listed in figure 5.1A. We will argue that females are able to control the cost of mate choice and of associating with males by adjusting their overall degree

A. Components of mating preferences

Mating preferences
Receptivity and responsiveness
Choosiness
Preference function

B. Plasticity in mating preferences

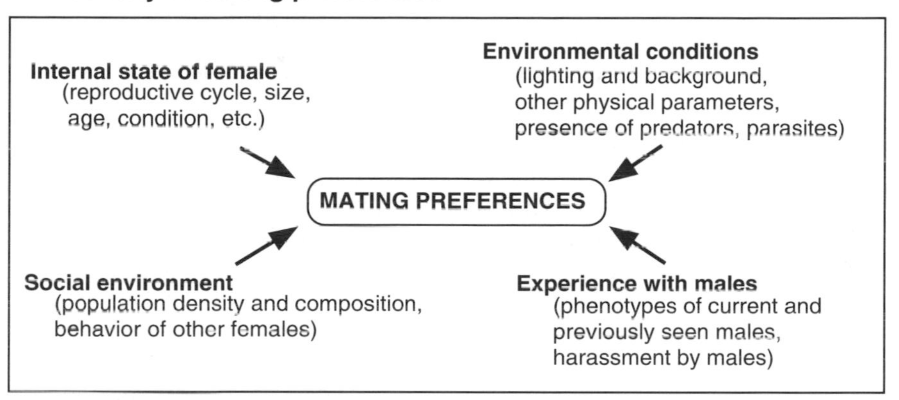

C. Selection on mating preferences

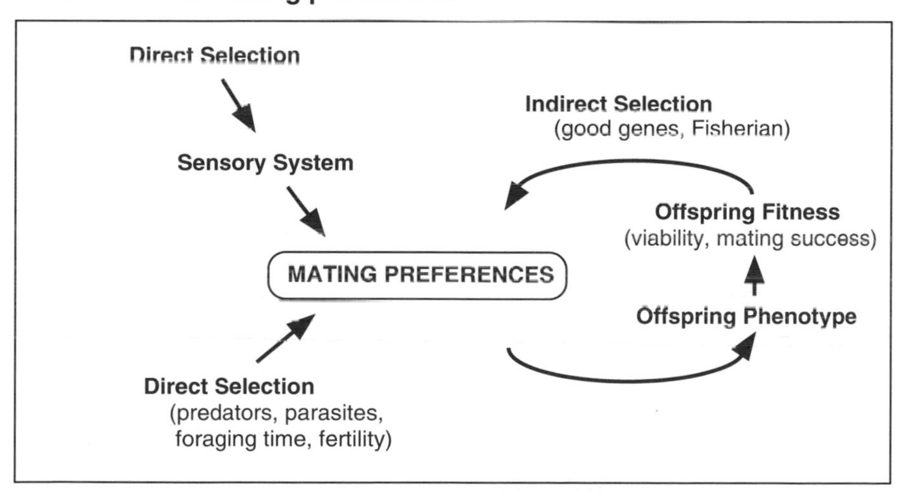

Figure 5.1 Plasticity and selection effects on female mating preferences behavior.

of responsiveness to males and the degree to which they actively avoid the attentions of males. Mate choice also depends on the degree to which females respond differently to different males, in other words, on their choosiness. Females appear to adjust their choosiness in response to costs of associating with males. Finally, the pattern of mating preferences can be defined by the specific relationship between male characteristics and female responsiveness, or a preference function (e.g., Lande 1981). In this chapter and the next, we will see that all of these elements may vary within and among individuals and among populations. Levels of female responsiveness and choosiness and the shape of preference curves can all affect the outcome of sexual selection and can contribute to divergence in male secondary sexual characters between populations.

5.3 Selection and the Evolution of Mating Preferences

Current sexual selection theory (e.g., Kirkpatrick and Ryan 1991; Schluter and Price 1993; Andersson 1994; Iwasa and Pomiankowski 1995) describes two major ways in which selection can affect the evolution of mating preferences (fig. 5.1C). Direct selection on preferences can occur when benefits or costs of mate choice have direct effects on the fitness of the female, through her survival, fecundity, future reproduction, or the survival of the offspring. Paternal care, resources provided or controlled by the male, risks of mate choice associated with predators or parasites, and time constraints could all result in direct selection on mating preferences. Sensory processes can mediate the effects of direct selection on mating preferences through "sensory drive" (see sec. 5.5 for discussion and references). Selection can also affect the evolution of mating preferences indirectly. This occurs when mate choice by females leads to fitness differences in offspring that result from characteristics inherited from males and includes the much-discussed "good genes" and "Fisherian" models of evolution by sexual selection (see chapter 6). This chapter examines the proximate mechanisms of mate choice and how they are shaped by the effects of direct selection.

Given that the mating system of guppies does not involve resources or oviposition sites controlled by males or parental care by males, there are relatively few ways in which females can obtain benefits of mate choice that directly affect their survival or fecundity. One possible direct benefit of female choice in guppies is the possibility of avoiding becoming infected by parasites during mating. Much has been made of Hamilton and Zuk's (1982) suggestion that mate choice can result in offspring that are genetically more resistant to parasite infection (see chapter 6). Relatively little attention has been paid to the simpler and more direct idea that individuals can reduce the chance of being infected themselves by assessing the health of potential mates (see Borgia and Collis 1989, 1990; Rosenqvist and Johansson 1995). Males infected with *Gyrodactylus* (see chapter 1) have reduced brightness of orange spots and reduced attractiveness to females (see chapter 6 for details). They may also be more likely to transmit parasites to a female during a copulation than a healthy male. Female guppies exposed to males with moderate *Gyrodactylus* infections were more likely to become infected if they copulated with the male than if they did not copulate (Bridges and Houde, unpublished data), but there was no difference in infection rate of females copulating with either heavily or lightly infected males.

Another possibility for a material benefit of mate choice is a nutritional contribution in the male ejaculate provided along with sperm (e.g., Markow 1988). Females might be able to judge the likely quality of a male's ejaculate and thus improve their nutritional state through mate choice. Only a minute amount of material is passed to females by male guppies, and a nutritional benefit is perhaps unlikely, but this needs to be examined in more detail.

Consideration of the possible costs of mate choice has led to a renewed interest in the role of direct selection in the evolution of mating preference (e.g., Pomiankowski 1987; Real 1990; Pomiankowski et al. 1991; Reynolds and Gross 1990; Crowley et al. 1991; Kirkpatrick and Ryan 1991; Price et al. 1993). Costs of mating, choosing, or associating with males can involve increased risk attack by predators or competitors (e.g., Borgia 1981; Reynolds and Côté 1995), increased risk of parasite transmission (e.g., Borgia and Collis 1989, 1990; Rosenqvist and Johansson 1995), reduced time available for foraging (Magurran and Seghers 1994a,c), and energetic costs of moving between males (e.g., Milinski and Bakker 1992). General reviews of costs associated with mate choice include Pomiankowski (1987), Gwynne (1989), Lima and Dill (1990), Reynolds and Gross (1990), Magnhagen (1991), and Sih (1994). In guppies, energetic costs of mate selection are probably low, because each female is approached by many males, rather than vice versa. Sexual contact may also involve a risk of physical damage for female guppies. I have occasionally observed internal bleeding and even death of females following mating. If these kinds of costs are important, we can view mate choice by females as a foraging problem, as we did with male mate-searching behavior. From the point of view of females, the problem is analogous to diet choice. Males vary in quality, and the "goal" for females has been modeled in terms of maximizing the quality of mates subject to constraints imposed by costs of mate choice (Janetos 1980; Real 1990, 1991; Wiegmann et al. 1996).

5.4 Mate Choice Patterns

Ignoring and Avoiding Males

The incessant attentions of males present a fundamental problem for females. Although copulation is essential for reproduction, from the females' point of view the courtship and copulation attempts of males are excessive and may even reduce the females' fitness, for example by reducing time available for foraging (Magurran and Seghers 1994a). Therefore, "harassment" by males may be a significant problem for female guppies, and females may benefit from avoiding males much of the time (Magurran and Nowak 1991; Magurran and Seghers 1994a,c). Females may be able to avoid males by moving to deep waters in some streams or by hiding under leaves and other objects (Reynolds, pers. comm.).

Because males tend to switch their attentions away from unresponsive females (Hankes and Houde, unpublished data), becoming unresponsive may be an effective way of avoiding the risks of associating with males. The main means females use to minimize contact with males seems to be by limiting their responsiveness to courtship displays to brief receptive periods and by cooperating in only two or three copulations each time they are receptive (see chapter 2). This also suggests that there are no net benefits of mating more frequently than two or three times per receptive period. Possible benefits of multiple mating for female guppies (e.g., fertility effects) remain to be investigated but could be tested easily with simple experiments. Females also actively avoid sneak copulation attempts by males (Magurran and Nowak 1991). The presence of more than one male near a female may be especially likely to attract predators and may explain the tendency of females to flee when pursued by more than one male (see chapter 2).

The presence of predators can also affect the sexual responsiveness and avoidance behavior of females directly. In a laboratory study, previously responsive virgin females exposed to a predator became unresponsive to males more frequently than did control females that did not see a predator (Gong 1995). Looking at guppies in the field, Magurran and Seghers (1994c) found that female guppies in the wild avoid males more and yet still receive more gonopodial thrusts in high-predation (*Crenicichla*) localities than in low-predation (*Rivulus*) localities. They attributed this difference to a decrease in time spent foraging by males in *Crenicichla* localities and a consequent increase in time spent following females. Females may also be more vulnerable to unwanted attentions of males in *Crenicichla* localities because they devote more time and attention to antipredator behavior. Ultimately, differences in the level of "harassment" by males could constrain the ability of females to exercise their mating preferences, affecting the overall pattern of sexual selection (see chapter 6). It would also be interesting to know if females are more or less likely to avoid a male in a given instance depending on risk of predation. Although females in *Crenicichla* localities avoid males more often per unit time because of an increased rate of sneaky mating attempts, the female's extra attention given to avoiding predators might be expected to reduce her ability to avoid a given male. This suggests that, for males, the sneak copulation strategy could be most effective in high-predation localities as well as being the safest strategy.

Females thus may be able to reduce some of the costs involved in mate choice by limiting their responsiveness and even actively avoiding males at times. But assuming that there are in fact benefits to choosing mates rather than mating randomly, we can predict that the selectivity with which females choose mates (choosiness) should be affected by the risk of predation or other costs of mate choice.

Predation and Variation in Choosiness

We can define the degree of choosiness or strength of preference in terms of the variance in the responsiveness of females to males (Endler and Houde 1995). A choosy female may be strongly responsive to some males and likely to copulate with them, but unresponsive to other males and very unlikely to copulate with them. An unchoosy female's responsiveness is about the same to all males, as is her chance of copulating with them. The degree of choosiness directly affects the strength of sexual selection in a population.

Theory predicts costs associated with mating and mate choice should affect the degree of choosiness shown by females (Hubbell and Johnson 1987; Pomiankowski 1987; Real 1990; Crowley et al. 1991; Pomiankowski et al. 1991). The same reasoning also applies to male choice of females (see chapter 4; Deutsch and Reynolds 1995; Fitzpatrick et al. 1995), and the effect of costs on choosiness needs to be examined in males as well as females. Variation in female choosiness has been documented in only a few cases, including guppies (Forsgren 1992; Berglund 1993; Forsgren and Magnhagen 1993; Hedrick and Dill 1993; Endler and Houde 1995; Gong 1995; Godin and Briggs 1996).

The choosiness of female guppies varies both genetically and through facultative behavioral responses to environmental conditions. Both kinds of variation appear to represent adaptive responses to costs of choosing associated with the risk of predation. Very choosy females are likely to allow many males to court them before actually mating and incur more of the costs of associating with males (see above), while unchoosy females are more likely to mate more quickly, after being courted by fewer males, and incur fewer such costs. Another cost of choosiness, the time spent searching for a mate, is less likely to be very important in most guppy populations because of the high rate at which females are approached by males.

The possibility that predation risk may affect choosiness has been investigated by Gong (1995) and by Godin and Briggs (1996). The prediction is that females should reduce their choosiness in the presence of a predator, and that this response to predation risk should be most evident in populations that are subject to high predation risk. The first prediction was confirmed in both studies. Females from both low-predation and high-predation localities tended to associate with the brighter or more actively courting of two males in the absence of a predator. But in visual contact with a predator, females tended to eliminate or reduce their preference for the brighter male. The two studies differ in their support for the second prediction. In Gong's (1995) study, both high- (greenhouse stock) and low-predation (Paria) females changed their patterns of mate choice when exposed to the predator. Godin and Briggs (1996), on the other hand, found that high-predation (Quare) females changed their pattern of mate choice but that low-predation (Paria) females did not. It is not clear what differences in experimental design might account for the difference in results for Paria females.

Overall, these studies suggest that females can adaptively adjust their degree of choosiness in response to predation risk, and also that there may be genetic variation in this behavioral plasticity. The conclusions that can be drawn from the experiments are compelling but limited by the design. The case for adaptive adjustment of choice could be strengthened by evidence, first, that females are actually able to reduce their contact with males, especially conspicuous males, in the presence of predators, and second, that they are actually able to reduce their risk of being preyed upon by doing so.

Variation in mating preferences between populations is also consistent with the prediction that risk of predation may affect choosiness. In a comparative study of several guppy populations (Houde and Endler 1990; Endler and Houde 1995; see also chapter 6), the strength of mating preference (especially for orange coloration) tended to be negatively associated with risk of predation. For this study, we measured preferences of lab-reared females in the absence of predators, and differences among populations were likely to be genetically based. Females with the strongest preference tended to be from populations that experience low levels of predation, and vice versa, suggesting that the strength of preference may depend on the level of predation. However, we cannot say for sure whether strength of preference is directly affected by selection, or if it has evolved through indirect selection (see chapter 6) or as a result of other processes.

Mate Choice Rules

The actual mate choices made by females are the result not only of their overall level of choosiness as discussed above, but of the sequence of encounters with males and the rules by which females decide which of those males to choose. A number of theoretical studies (Janetos 1980; Parker 1983; Seger 1985; Real 1990, 1991; Wiegmann et al. 1996) have examined some of the behavioral rules that females could use in mate choice. These include choosing the best of a sample of *n* males ("best-of-*n*" rule), and sequential searches ("one-step" processes) in which there are fixed or relative thresholds for acceptance of males. When costs of sampling males are very low, and females are able to return to previously sampled males or to view several males simultaneously, the best-of-*n* rule appears to be best (Real 1990). When costs of searching and sampling are significant, a sequential search rule is generally better. The extreme case of this is when costs are very high and females are predicted to mate with the first male encountered.

Only a few studies have investigated the details of female choice rules. Female preferences involving single male tests (see Appendix) provide evidence that females have absolute internal standards (or thresholds) for mate choice that determine their responses to different males. This is because tests using single males and females that have no prior experience with males give females no opportunity to base their responses on comparisons among males. The possibility that simultaneous or sequential comparisons among males might alter this basic pattern of preference cannot be ruled out without further experiments, however. Examples in which absolute internal standards have been demonstrated include studies of cockroaches *Nauphoeta cinerea* (Moore and Moore 1988) and red jungle fowl (Zuk et al. 1990a), as well as guppies (Houde 1987; Reynolds and Gross 1992; Brooks and Caithness 1995a).

In addition to these absolute criteria for mating preferences, females may use relative choice rules in which they compare males. In sticklebacks, females sample males sequentially, and the phenotype of the previously sampled male affects the female's response to the subsequent male (Bakker and Milinski 1991; Milinski and Bakker 1992). Females spent less time with males with dull throat patches if the previous male had a bright throat patch than if the previous male had a dull throat patch. This pattern of sequential choice depends on search costs, however. When the costs of swimming against a current were high, females spent longer with dull males that were preceded by bright males (Milinski and Bakker 1992) than when costs were low.

Female guppies encounter males sequentially, as each male approaches, courts, and eventually moves on (chapter 2). Thus a simple sequential sampling process would seem to be a plausible initial model, but more complex models also need to be examined. The female's response could be an absolute function of traits such as the color pattern exhibited by that male, but it could also depend on the traits of previously encountered males or her prior experience (see next section). Nordell (1995 and pers. comm.) has found some evidence that the phenotype (orange area) of the first of several males sampled can affect how quickly the female moves on to the next male. In another study, Brooks and Caithness (1995a) looked at the responsiveness of females to each of two males, to which they were introduced sequentially, in separate tests. They showed that, although the responsiveness of females to the first male sampled was strongly dependent on orange area, there was no significant effect of orange area on responsiveness to the second male sampled. This suggests that the response to the second males may have been modified depending on the relative attractiveness of the two males. This inference is supported by the further result that the difference in the females' response to the first and second males was significantly correlated with the difference in orange area between the two males. These findings parallel the results Bakker and Milinski (1991) obtained with sticklebacks.

Unlike sequential sampling processes, "best-of-*n*" search rules envision a more extensive comparison among a subset of males in the population. In this case, the female returns to the male she prefers out of those sampled. This kind of sampling has been documented in cock-of-the-rock *Rupicola rupicola* (Trail and Adams 1989), a lek-breeding bird in which females visit the display sites of males. It is hard to imagine that female guppies are likely to go back systematically to previously sampled males in natural social situations, given that males approach females rather than vice versa, but it is a possibility worth exploring.

Effects of Prior Experience

The mate choice patterns of females can also be affected by the characteristics, behavior, and demography of conspecifics encountered prior to the current series of encounters. These kinds of effects on male behavior were studied by Rodd and Sokolowski (1995), but less work has been done on early influences on female choice. Breden et al. (1995) examined the possibility of such effects by exposing maturing females to particular colorful or noncolorful male phenotypes. In some but not all subsequent mate choice tests, females that had been reared with colorful males showed greater preference for (or less avoidance of) more colorful males than did females reared with noncolorful males. Experience with males during maturation, therefore, may affect the direction of preference as well as the degree of preference. The results of Breden et al. (1995) are suggestive of sexual imprinting (e.g., Bateson 1978), in which familiar individuals are often preferred as mates. Sexual imprinting is sometimes interpreted as a mechanism for achieving optimal outbreeding (Bateson 1983), although there may be other benefits of mating with familiar individuals. Breden et al. (1995) discuss possible consequences of the effect for population differentiation. The changes in patterns of mate choice in Breden et al.'s (1995) experiment can also be interpreted from the point of view of adaptive mate-searching strategies. Prior experience with male phenotypes may provide information about the expected quality of males in the population, allowing females to fine-tune their mate choice.

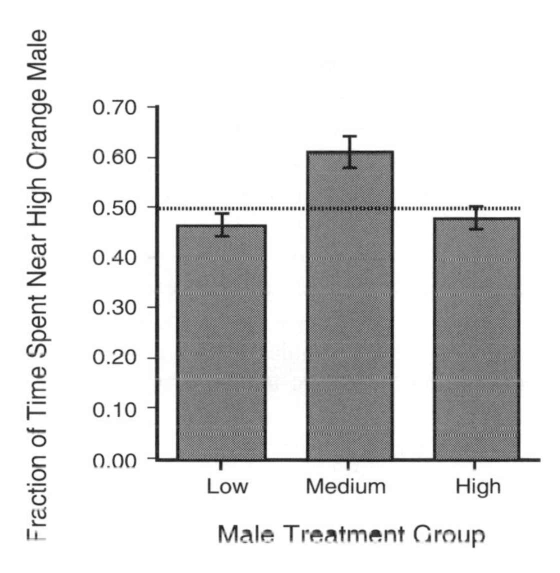

Figure 5.2 Mating preference of females reared with high, medium and low orange males in dichotomous-choice tests. Dotted line at 50% indicates equal time spent with high and low orange males. (From Rosenqvist and Houde, in press.)

In an experiment similar to that of Breden et al. (1995), Rosenqvist and Houde (in press) exposed females to groups of males with low mean orange area in their color patterns (low treatment), high mean orange area (high treatment), or a mixture of high-orange and low-orange males (mixed treatment). In preference tests, only the females from the mixed treatment showed a significant preference for high-orange males (fig. 5.2). Females that had seen only low-orange or only high-orange males showed no preference. These results also show that mate choice patterns can be modified depending on prior experience with male phenotypes, but are not consistent with a simple sexual imprinting model in which females prefer familiar males. Instead, it appears that females exposed during development only to highly attractive males (high treatment) or only to relatively unattractive males (low treatment) seem to reduce their choosiness. Females exposed to the mixed treatment saw both attractive and unattractive males, and continued to be choosy. The key difference between the treatments appears to be the information about male quality provided by orange coloration—relatively little information was provided in the low and high treatments, while more information was provided in the mixed treatment. We can argue that in the absence of information, females may be reducing their costs by reducing their level of choosiness.

Copying

Mate choice may also be modified by females' interactions with other females. The idea that females may observe the mate choices of other females and then mate with males chosen by other females (Losey et al. 1986; Gibson and Hoglund 1992; Pruett-Jones 1992; Kirkpatrick and Dugatkin 1994) was applied to guppies by Dugatkin (1992a), and has been confirmed in further experiments by Dugatkin and his coworkers. An independent test for copying by Brooks (in press) suggests that some but not all females may reverse their mate choice when given the opportunity to copy.

Dugatkin (1992a) allowed females to observe two males, one of which had another female (the model female) placed next to him. He found that females then preferred the male that had been accompanied by the model female once the model female was removed and concluded that the focal females were using the proximity of the model female as a cue for mate choice. He was able to rule out several alternative explanations for his copying result using clever variations on the basic experimental design, and later showed that the presence of a model female could actually reverse an already established preference (Dugatkin and Godin 1992b). A subsequent study (Dugatkin and Godin 1993) revealed that younger females are more likely to copy the mate choice of older females than vice versa. This suggests that older females may be "better informed" about appropriate mate choice decisions than younger females and/or that younger females have more to gain through copying than do older females. Copying has also been documented using the same experimental design in sailfin mollies and Japanese medaka *Oryzias latipes* (Schlupp et al. 1994; Grant and Green 1996).

In another experiment, Briggs et al. (1996) attempted to test the effect of the presence of a predator on the tendency of females to reverse their mate choice when given the opportunity to copy. This study further confirmed that female guppies tend to reverse their mate choice when they can copy the mate choice of another female. The frequency of reversals was similar in the presence of the predator to that with the predator absent. This suggests that predation risk may have no effect on the tendency to copy; but it is also possible that in the presence of the predators, females were reversing their mate choice as a direct response to the predator rather than because of the opportunity to copy. Additional experiments will be needed to sort out these alternatives.

Taken together, experimental results provide good evidence for copying. However, the role of copying in freely interacting social groups of guppies remains unclear. Especially worthwhile would be independent evidence for copying using a different experimental design. There is one further alternative explanation of apparent copying that also needs to be investigated. This is that females may be choosing males previously seen associating with another female, not because this association is an indication of the male's attractiveness, but because the association indicates that the male is capable of monopolizing females, possibly an advantageous characteristic in a mate especially if harassment by other males is a problem (Magurran and Nowak 1991; Magurran and Seghers 1994c; see above). In either case, the effect is to modify patterns of female choice, which may in turn affect the dynamics of sexual selection in populations (Wade and Pruett-Jones 1990).

Dugatkin (1996) has extended his work to determine the relative importance of social factors (copying) and genetic factors (color pattern differences) in mate choice by female guppies. Given that we have evidence that female choice probably contributes to the evolution of color patterns, including orange spots in guppies, to what extent does copying tend to reverse the effects of females' intrinsic preferences? Dugatkin (1996) found that discrimination between small (12%) and moderate (24%) differences in orange area could be reversed by copying, but that when two males had larger (40%) differences in orange area, females maintained their preference for the more orange male even when given the opportunity to copy a female associating with the less orange male. We can conclude that there should still be a net effect of the females' intrinsic preferences on the evolution of orange coloration, even though copying may reduce the intensity of sexual selection.

5.5 Sensory Mechanisms

The preceding discussion has focused on ways in which various factors affect how a female chooses her mate, or her mate choice strategy. Mate choice also depends on how females perceive males, and the effects of selection on mating preferences may depend on the sensory processes involved in perception of potential mates (e.g., Endler and McLellan 1988; Ryan 1990; Endler 1992; Ryan and Keddy-Hector 1992). To fully understand how mate choice comes about and how it evolves, we must understand how females process information about male phenotypes.

A difference in the way a female responds to males that differ in color pattern implies a sensory or cognitive bias in the female's perception of the males at a proximate level. Investigation of the sensory mechanisms behind female choice has begun only recently. This approach has been fruitful for systems in which choice is based upon characters that are relatively simple in terms of the nature of the signals sent and received. Work with frogs and their auditory systems, for example (e.g., Ryan and Rand 1990; Wilczinski et al. 1995), has revealed that female preferences for particular aspects of calls is based on the pattern of sensitivity to particular sound frequencies.

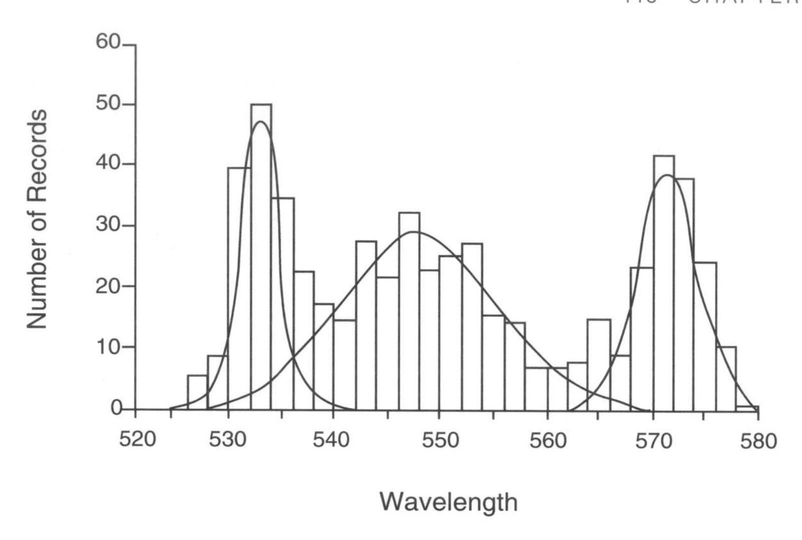

Figure 5.3 Polymorphism in spectral sensitivity of guppy visual pigments. The graph shows a frequency histogram of the maximum wavelength absorbance (λ_{max}) of individual long-wavelength cone cells from the retinas of guppies. Records from several cells from each of fifteen guppies were combined. Normal curves fitted to the data suggest that there are three types of long-wavelength cones, with maximum absorbance estimated at 533, 548, and 572 nm, respectively. (Adapted from Archer and Lythgoe 1990.)

Similarly, in guppies, we can investigate the sensitivity of the retina to particular light frequencies in order to learn about how females perceive males. Given that females from some populations do respond preferentially to males depending on characteristics of color patterns, there must be a bias in the processing of color pattern information somewhere in the visual or cognitive systems of female guppies. Similarly, genetic differences in mating preferences may be the result of differences in visual processing at some level. At the most simplistic level, preferences for particular color pattern characters could be determined by sensitivity to corresponding light frequencies, and differences among populations could be the result of differences in visual pigments, retinal cell populations or other features of the female's visual system. In sticklebacks, for example, changes in visual sensitivity are correlated with seasonal changes in breeding status of females (Cronly-Dillon and Sharma 1968). This suggests a possible role for visual sensitivity in mate choice. In guppies, individual variation in the peak absorbance of long-wavelength visual pigments, may lead to differences in vision in the red part of the spectrum (Archer et al. 1987; Archer and Lythgoe 1990; fig. 5.3), as a result of individual variation in sensitivity of the retina to red light. Artificial selection experiments suggest that the variation in sensitivity to red and blue light is heritable (Endler, pers. comm.). This kind of variation could form the basis for differences in

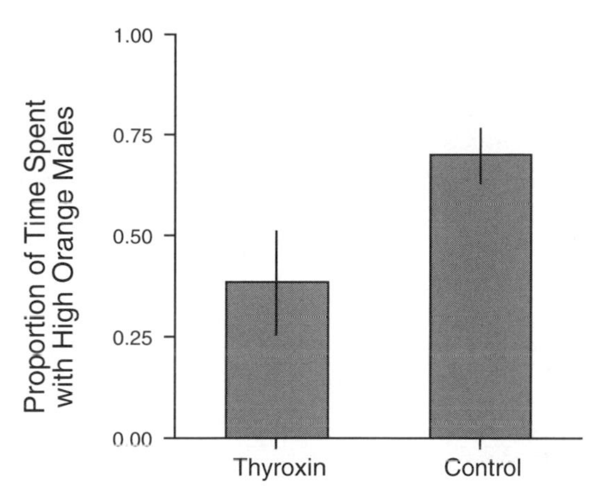

Figure 5.4 Effect of thyroxin treatment on mating preference of female guppies. Error bars indicate one standard error. (Adapted from Rush, in press.)

mating preferences between guppy populations, but this remains to be demonstrated.

Hints that differences in mating preference are related to basic functions of the visual system come from physiological manipulation of the visual system (Rush, in press) and from an artificial-selection experiment described in more detail in chapter 6 (Houde 1994). Rush (in press) treated female guppies with thyroxin, which alters the spectral sensitivity of visual pigments. Treated females did indeed show a shift in their spectral sensitivity as measured by electroretinography, becoming more sensitive to orange-red light. In subsequent mate choice tests, these females proved to be less discriminating based on the amount of orange in male color patterns, relative to untreated control females (fig. 5.4). Initially this result seemed surprising—one might expect to find greater sensitivity to orange light in fish that show preferences for orange in color patterns. Apparently, fish with high sensitivity to orange can easily see even very small amounts of orange in a male's color pattern, and are therefore responsive to most males. On the other hand, fish with lower sensitivity may be able to detect or respond only to males with large amounts of orange in the color pattern and are therefore more discriminating.

In another experiment (see chapter 6 for more details), selection for increased and decreased amounts of orange in male color patterns led to a corresponding shift in female preferences, resulting in females with high and low expression of preference for orange. The spectral sensitivity of selection lines with high and low preference was then tested using electroretinography as in the thyroxin experiment (Houde and Rush, unpublished data). Low-preference females were more sensitive to orange-red wavelengths and also to blue-violet wavelengths than were high-preference females, a result that exactly corresponds to the findings from the thyroxin experiment described above. It appears, therefore, that change in patterns of mate choice may depend on very simple changes in the spectral sensitivity of the visual system. Further evidence is needed to confirm this, however.

It is also possible that selection could affect sensory functions in a context other than mate choice. This could result in sensory biases leading to new mating preferences or to evolutionary changes in existing mating preferences (Basolo 1990a). For example, selection of increased efficiency in detecting a particular food item could lead to increased sensitivity to particular colors or shapes. This in turn could bias the responsiveness of females to particular traits displayed by males. Recent observations that female guppies are especially likely to react to red food items (F. H. Rodd, pers. comm.) suggests that this argument may not be as far-fetched as it might seem.

Another proximate way in which female choice can be modified is if the visual environment changes the way in which females perceive males. The perception of males by females depends on the interplay between the female's sensory system and physical factors in the environment that affect the transmission and reflection of light (Endler 1991). Interspecific variation in the color patterns of birds, for example, has been related to variation in the visual environment (Marchetti 1993; Price 1996). Geographic variation in calls of frogs appears to be related to variation in the auditory environment (Ryan and Rand 1990). For fish, the light transmission properties of water can affect the spectrum of incident light when a female views a male's color pattern. Differences in incident light can have dramatic effects on the appearance of color patterns (Lythgoe 1979; Endler 1978, 1990, 1991, 1992, 1993) and can change the pattern of preferences shown by individual females (Long and Houde 1989; see discussion in chapter 6). Effects of water color may contribute to differences in color patterns among populations of sticklebacks (Reimchen 1989) and guppies (Endler and Houde 1995; see discussion below in this section).

Evolutionary changes in communication systems resulting from how signals are transmitted and perceived have been termed "sensory drive" (Ryan et al. 1990b; Endler and McLellan 1988; Endler 1992; Schluter and Price 1993). Sensory drive refers to the effects of sensory mechanisms on the evolution of signal-receiver systems. Sensory drive models predict the evolution of male traits and female preferences for traits with maximum "detectability" to females in a given environment (Schluter and Price 1993). These are essentially direct selective effects on mating preferences. Preferences for the most detectable male traits have advantage of greater efficiency and reduced error rate in mate choice. Preferences that minimize

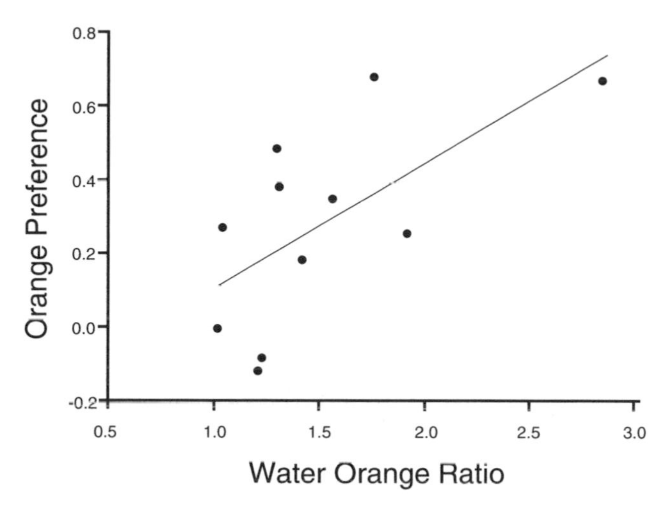

Figure 5.5 Relationship of mating preference of females to orange ratio of the water in population's native stream. Eleven populations are compared. (Adapted from Endler and Houde 1995.)

contact with males that are conspicuous to predators would also be advantageous. In guppies, we would then predict the evolution of preferences (and associated color patterns) that maximize perceived conspicuousness of males to females, minimize conspicuousness of the courting pair to predators, and also minimize the time spent in mate choice by females. The way females perceive male color patterns depends on the visual environment (spectrum of incident light and transmission properties of water), the spectral sensitivity of predators, and the spectral sensitivity of females (Endler 1992; Endler and Houde 1995). Thus, variation in the visual environment and in predator species can lead to different selection regimes for the sensory processes underlying female preferences, possibly accounting for some of the variation in preferences among populations (Endler 1991, 1992, 1993).

The evidence that sensory drive processes may affect the evolution of mating preferences and color patterns in guppies is indirect at best (Endler and Houde 1995). For example, variation in water color may contribute to divergence in male color patterns and female preferences among guppy populations (fig. 5.5; Endler and Houde 1995). Also, in turbid water, male guppies appear to switch their mating tactic from primarily display to primarily sneak copulation (Luyten and Liley 1985). This suggests that female mating preferences may emphasize visual cues less in turbid waters where visibility is poor than in clear waters where visibility is good. This prediction remains to be tested, however. In stickleback populations, water color may be responsible for differences in the throat color of males, but effects on female preferences are not clear (McPhail 1969; Reimchen 1989; McKinnon 1995; see further discussion in chapter 6).

In comparing guppy populations, Endler and Houde (1995) found that preferences for orange were strongest in populations from tea-stained streams with greater transmission of long (orange-red) than short (yellow-green-blue) wavelength light (high "orange ratio"; see fig. 5.5). Weak orange preferences tended to occur in populations from streams with more equal transmission of long and short wavelength light (orange ratio near 1.0). Reimchen (1989) found the opposite pattern with sticklebacks: in tea-stained streams, males had black throat patches and females did not prefer males with red patches, while in streams with uncolored water (orange ratio near 1.0), males had red throat patches and females preferred males with red patches. A possible explanation for this discrepancy is that in both cases, females prefer the most detectable males, but that a difference in the visual background affects which males are most detectable (Endler and Houde 1995). In guppies, males are viewed against a visual background of stream bottom, against which orange spots are expected to be especially conspicuous when the water is tea stained. In sticklebacks, streams are often deeper, so males are viewed with no stream-bottom background, and tea-stained water is expected to make red throat patches appear relatively inconspicuous. Obviously, these arguments are in need of further substantiation.

Also consistent with sensory drive predictions, variation in predation regime tends to predict which particular color pattern elements are preferred or not preferred by females (see Endler and Houde 1995 for details). In many but not all cases, females from low-predation populations prefer colors that are most conspicuous to them under the light conditions when courtship occurs, while females from high-predation populations tend not to prefer colors that are most conspicuous to predators. Thus, differences in preferences among populations may depend on interactions between perception of color patterns by females and perception of color pattern by predators, which in turn depend on lighting conditions at the times of the day when courtship and predation are most likely to occur (Endler 1991, 1992, 1993).

Although Endler (1991, 1992, 1993) has convincingly argued that male color pattern elements that are conspicuous to females are not necessarily those that are conspicuous to predators, color pattern elements may often be conspicuous to both. In order to further evaluate the sensory drive idea, we will need to weigh the prediction that females which avoid dangerously conspicuous males can reduce their likelihood of being preyed upon against the prediction that females which prefer highly detectable males benefit because they minimize the time they spend in risky interactions with males. Both of these predictions need to be tested more thoroughly.

5.6 Summary

Despite the difficulties inherent in finding out what females "think" about males, studies of guppies have provided valuable insights into the proximate mechanisms and patterns of variation in female mating preferences. Much of this information is without parallel and needs further investigation and corroboration in other species.

The general picture that emerges from the guppy studies is that mate choice, like so many other characteristics of guppies, is shaped by predation. Time and energy constraints do not appear to be particularly important in shaping mate choice in guppies, but all aspects of mating and mate choice are likely to involve costs associated with the risk of predation. Female guppies appear to minimize the amount of time they spend responding to males by limiting receptive periods to only a few days per cycle and by actively avoiding males at other times. In addition to costs associated with predation, other possible costs of contact with males (for example, parasite and disease transmission) need to be investigated. Females adjust their responses to males in more subtle ways as well when predators are present. Individual females can adjust their degree of choosiness among males in response to a predator, and population levels of choosiness may depend on predation regime.

Although choosiness directly affects the outcome of sexual selection on male characters, the details of how females choose their mates are also of interest but are less well understood. Again, costs associated with predation may ultimately determine the rules females use in mate choice. Mate choice rules describe how the present sequence of encounters with males results in mate choice. At present we do not have a detailed understanding of mate choice rules in any species, including guppies, although work with sticklebacks suggests avenues for further study. We do have some information showing that mating preferences of female guppies can be modified depending on prior experience, although further work is needed to understand how females integrate both long-term and short-term experience into their pattern of mate choice. Guppies provide the best example of mate choice copying in which a female will base her mating preference on prior interactions she has observed between males and females. More information on copying is now needed from species other than guppies. In addition, females seem to modify their mate choice based on the phenotypes of males they have experienced in the past. Both copying and other effects of prior experience can be interpreted as adaptive flexibility that may help minimize costs involved in associating with males and mating, but specific adaptive hypotheses have yet to be tested.

Finally, guppies, along with only a few other species, have been used to explore the proximate sensory basis of mating preferences. Evidence is mounting that simple changes in visual sensitivity, perhaps at the level of the photopigments themselves, can account for observed differences in mating preferences. Furthermore, we are realizing that the perception of males by females (or predators) depends on the physics of signal transmission, which in turn depends on environmental variables such as forest vegetation and water color. We have scanty evidence suggesting that interactions between the sensory systems of females and predators and physical factors in the environment can affect the evolution of mating preferences of females and secondary sexual traits of males.

There are clear indications that individual female guppies adjust their patterns of mate choice in adaptive ways and that variation in mate choice between populations may also represent adaptive responses to direct selection. There are thus striking parallels between patterns of variation in male sexual behavior (fig. 4.1) and in female mating preferences (fig. 5.1). The overall picture, however, is sketchy, and more work on all aspects of female choice is needed to fill in the details for guppies and also for other species. One theme that has arisen repeatedly is that the ability of females to express advantageous mating preferences is constrained by the risk of predation and by the behavior of males (sneak copulations). Additional studies of how these constraints on female choice affect the evolutionary outcome of sexual selection should be especially interesting. In particular, the degree to which female preferences and male characters coevolve in parallel, as described in the next chapter, may depend on the degree to which females are constrained in their expression of mating preferences.

6 Evolution of Female Choice 2: Indirect Selection, Variation, and Correlations

6.1 Indirect Selection and the Evolution of Mating Preferences

Much of the focus of work on sexual selection has been on ultimate explanations for the evolution of both female choice and male secondary sexual characters. In chapter 5, we examined proximate ways in which patterns of female mate choice in guppies are shaped, for example by the direct costs of associating with males. But an ultimate explanation for the origin and maintenance of the mating preferences of female guppies and the color patterns of male guppies requires us to consider the possibility that traits inherited by offspring from their fathers and mothers can lead to indirect selection on mating preferences. This chapter examines the evidence from guppies that mating preferences evolve because of effects on general components of fitness of offspring ("good genes" models), and because of effects on the mating success of sons ("Fisherian" models).

Guppies provide a classic example of a non-resource-based mating system (e.g., Borgia 1979) in which males provide no resources or parental care to females nor do they defend either resources needed by females or females themselves. Thus, resources and benefits obtainable from males initially did not seem likely to provide direct advantages for mate choice based on color patterns in guppies (but see chapter 5 for the role of direct costs of mate choice). Much of the work on mating preferences in guppies has therefore focused on examining models of indirect rather than direct selection.

The guppy system offers us one of the best opportunities for exploring some of the theoretical issues concerning the evolutionary dynamics of sexual selection. Guppies are well suited for examining hypotheses about inherited effects of female choice. The possibility of carrying out breeding experiments with guppies over several generations and the fact that color patterns are heritable and genetically well understood make this system ideal for getting at some of the thornier issues.

6.2 Heritable Variation in Mating Preferences

Most models of the sexual selection process involve evolutionary changes in both male traits and female mating preferences. Thus preferences are implicitly assumed to have heritable genetic variation. Given that this is a departure from earlier ideas of fixed preference patterns, it is of interest to know whether this heritability does indeed exist. On the other hand, common wisdom (Falconer 1989) says that almost any measurable trait is likely to have some additive genetic variance. The problem is the difficulty in measuring mating preferences accurately enough to estimate heritability or even to test for significant heritability within populations. Nevertheless, heritable genetic variation in mating preferences has been demonstrated in a number of species (see reviews by Bakker 1990 and Bakker and Pomiankowski 1995).

Several lines of evidence confirm that there is likely to be additive genetic variance for mating preferences in guppies, although heritability estimates using standard quantitative genetic methods have not been obtained. First, female guppies appear to be consistent in their preferences and to differ individually from one another. In an early study I found significant differences among females in their relative responsiveness to different males (Houde, unpublished). Godin and Dugatkin (1995) performed a formal repeatability analysis for mating preferences of female guppies presented with a choice of two males. They calculated repeatability to be highly significant at 58%. Brooks (in press) also found significant repeatability of mate choice in feral South African guppies. Repeatability of mating preference is high in some other systems (sticklebacks: Bakker 1993; barn swallows, *Hirundo rustica*: Møller 1994a), but low and nonsignificant in others (e.g., Boake 1989). Repeatability values are usually interpreted as representing an upper limit on heritability for that trait, although low repeatability measurements may be more an indication of the difficulty of measuring mating preference than of a constraint on evolutionary potential of underlying traits. Finding that repeatability is significantly greater than zero is an indication that we are able to measure mating preferences in meaningful ways, and that heritable variation in preferences is at least plausible.

Second, mating preferences diverged over a few generations in an artificial selection experiment (Houde 1994; see below for details of this experiment), another indication of heritable variation. In this case, I was not able to estimate heritability from the response to selection. Because this was an indirect response to artificial selection on male color patterns, the selection differential on preferences could not be calculated. The divergence in preferences does imply that the base population had additive genetic variance for this trait, however.

Finally, comparisons of preferences between populations show that there are genetic differences in mating preferences for orange coloration and other color pattern elements (Houde 1988a; Houde and Endler 1990; Endler and Houde 1995; see below for more details). Differences among populations confirm that preferences have indeed undergone evolutionary divergence in nature. But what are the mechanisms that lead to the evolution of mating preferences?

6.3 Good-Genes and Fisherian Models

The Fisherian and good-genes models of sexual selection have often been discussed as alternative explanations for the evolution of female mating preferences and the consequent evolution of male secondary sexual characters. Good-genes and Fisherian models for the evolution of mating preferences both involve indirect rather than direct selection on preferences. Both involve selection on characters that become correlated with preferences in the offspring resulting from mate choice.

Consider a population in which females have a mating preference for a particular male character. If both the mating preference and the male character have heritable genetic variation, then matings will be nonrandom with respect to the two characters. Females with a particular preference will mate with males with the corresponding expression of the preferred character, resulting in a genetic correlation between the two traits in the offspring. Tail length in birds (or guppies) is a standard example. In our hypothetical population, females vary genetically in the length of tail they prefer in their mates: some females prefer males with relatively long tails and others prefer shorter-tailed males. Females that prefer long tails end up mating with males with long tails, and females that prefer short tails mate with short-tailed males. Offspring of these matings tend to carry genetic information for corresponding values of both traits: long-tailed sons carry genes for the preference for long tails, and so on. In addition, any traits that are genetically correlated with tail length in males also become correlated with the mating preference in the offspring. This is important for good-genes models.

The significance of the genetic correlation between the mating preference and the male character is well known to quantitative geneticists. Selection on one of the characters results in evolutionary change in both characters. In Fisher's (1930b) model, if natural selection acts on the preferred character in males, resulting in adaptive evolutionary change, then the mating preference of females is predicted to undergo correlated, nonadaptive change. The change in the mating preference of females is considered "nonadaptive" because it is simply the indirect, correlated consequence of direct selection on the male character. The nonadaptive nature of the evolution of female preferences is one of the main points of debate concerning the Fisherian model of sexual selection. For many, the idea of adaptive evolution of mate choice is a much more intuitive and satisfying concept.

According to Fisher's (1930b) model, the correlated changes in female mating preferences can then result in further change in the average male character preferred by females. In the tail length example, natural selection for a slight increase in tail length leads to a correlated increase in the length of tails preferred by females. This results in sexual selection by female choice, favoring further increases in tail length and further correlated changes in female mating preference. Thus we have a self-reinforcing system of evolution of the two traits, female preference and male character, by sexual selection.

In good-genes models, by contrast, preferred male characters are genetically correlated with other viability-enhancing traits (see further discussion in secs. 6.3 and 6.4). When these viability traits are inherited by offspring and are subject to selection, the female preference is subject to further indirect selection. The evolution of preferences through the good-genes process is considered "adaptive" in that it is the result of effects on components of offspring fitness other than just mating success. Since the viability traits ("good genes") are correlated with the preferred traits, the Fisherian process as well as the good-genes process can contribute to the evolution of preferences simultaneously. The good-genes effect thus adds on to the Fisherian effect as a result of the correlated viability traits. The Fisherian model has therefore been viewed as a "null" model for sexual selection (Bradbury and Andersson 1987; Kirkpatrick and Ryan 1991).

Efforts to model the system explicitly (e.g., Lande 1981) have confirmed Fisher's (1930) claim that the cycle of sexual selection could end up as a "runaway" process in which evolutionary rates of both traits increase exponentially and lead to extreme exaggeration of the male trait. The key is the degree of genetic correlation that is established between the preference of females and the preferred character in males. If the correlation is strong enough, then the correlated change in female preference when the male character changes will be great enough to produce an even greater change in the male character in the next generation as a result of sexual selection. Alternatively, if the degree of genetic correlation is less, the two traits are predicted to evolve to an equilibrium at which the effect of sexual selection by female choice is exactly balanced by the effect of natural selection on the male character. For a given system, there are theoretically an infinite number of equilibrium points forming the now famous "line of equilibria" of Lande (1981) and Kirkpatrick (1982). Depending on initial conditions, a population may evolve to a high average expression of the male trait and corresponding female preference, a low average expression of both traits, or any point in between.

Although this Fisherian, or "line-of-equilibria" model is now generally accepted as a basic underpinning of sexual selection theory, it has been notoriously difficult to obtain relevant empirical data. Some unanswered questions include: (1) Do mating preferences actually evolve as a correlated effect as predicted? (2) Are there additional factors (direct, pleiotropic, or good-genes selection) that affect the evolution of female mating preferences? and (3) Are there examples of sexual selection leading to a runaway process, or do sexual selection processes generally lead to a stable evolutionary equilibrium?

Guppies may provide some answers to these questions in the future, but at present the data are incomplete, as they are for other study systems. Although a major research program has attempted to find ways to discriminate between good-genes, Fisherian, and other models for the evolution of female mating preferences (e.g., Bradbury and Andersson 1987), the general conclusion that appears to be emerging from studies of guppies and other species is that no one model is likely to be wholly correct or to apply in the absence of other models (Endler and Houde 1995). Instead, several mechanisms may simultaneously affect the evolution of mate choice in guppies, and different mechanisms may operate in different populations under different conditions. We have already looked at the evidence for effects of direct selection and sensory drive on the evolution of mating preferences in chapter 5, and we will now focus on the evidence for good genes and Fisherian processes.

6.4 Mate Choice for "Good Genes"?

Good-genes models are intuitively appealing because they provide an adaptive mechanism for the evolution of mating preferences. Selection on mating preferences is indirect, however, because neither the survival nor the reproductive success of females themselves is directly affected. The basic argument is that female preferences function to identify males with high "genetic quality." A preference for a particular character in males is favored if it results in an increase in fitness (e.g., viability) of offspring as a result of genes inherited from the father. Or, as stated in explicit mathematical formulations, the mating preference evolves because it becomes genetically correlated with genes that confer increased viability or other aspects of fitness in offspring. The genetic correlation arises because offspring inherit the preference as well as the good genes. An increase in offspring fitness leads to indirect selection on the mating preference, which in turn favors the male character on which the preference is based. In order to make a case that female choice could function to identify heritable good genes in males, we must demonstrate that the character on which choice is based is related to the quality or condition of males. More rigorously, the male character must be genetically correlated with (or predictive of) offspring fitness. The sometimes highly speculative hypotheses that attempt to link the sexual displays of males to offspring fitness have also contributed to the skepticism, controversy, and debate about good genes and handicap models.

Another difficulty with good-genes models is that intuitively plausible verbal models have often not worked when analyzed mathematically or with computer simulations. In an early attempt to explain the evolution of preferences for male characters, Zahavi (1975) argued that secondary sexual traits of males function as handicaps. Females that mate with males with extreme expression of the handicap obtain good genes for their offspring because males must be healthy and vigorous to survive despite the handicap. This corresponds to the "fixed" handicap model of Maynard Smith (1991). As it turned out, explicit versions of Zahavi's handicap model failed to predict the spread of the female preference and the male trait in populations mainly because of the negative fitness effects of the handicapping trait. The basic argument for guppies, according to the handicap model, would be something like the following. The color patterns of male guppies are conspicuous and increase the risk of predation and hence can be thought of as handicaps. Males that can survive despite having highly conspicuous color patterns must be superior in some way. Females that choose these males are therefore likely to have superior offspring, so their mating preference should be favored. Empirically, the handicap model for guppies would hinge on heritable variation in the ability of males to escape from predators. This is testable in principle, but other far more plausible adaptive scenarios for the evolution of mate choice based on color patterns have been proposed.

It is only very recently that tenable versions of good-genes models have been produced (Pomiankowski 1988; Grafen 1990a,b; Iwasa et al. 1991; Pomiankowski et al. 1991). The models that work best are those that assume that the degree of expression of the sexual character depends on the male's condition, i.e., that the fitness costs of producing the sexual character are condition dependent. Only males in good condition are able to show extreme development of the sexual character: the character must be more costly for a male in poor condition to produce and relatively cheap for a male in good condition to produce. Carotenoid coloration and courtship displays of males could be examples of condition-dependent characters in guppies (see next section).

In guppies and other species, two kinds of data can provide evidence for a good-genes model of mate choice based on condition-dependent characters. In the more proximate kind of approach, some studies attempt to demonstrate that the expression of sexual characters is indeed condition dependent. But ultimately, we need to show that males with preferred characters have in some sense fitter offspring than unpreferred males.

Condition Dependence of Male Traits

The fact that carotenoid pigments are important in mate choice in guppies (see chapter 3) suggests that this kind of argument based on condition dependence could apply. Carotenoids are obtained only from the diet and cannot be synthesized (see Endler 1980, Kodric-Brown 1989, and Lozano 1994 for literature reviews on carotenoids). This fact links the expression of carotenoids in color patterns to aspects of the physiological health of individuals through nutrition. In addition, it is common lore among aquarists and bird breeders that the brightness of colors, especially carotenoids, are indicative of the general level of health and vigor of individuals (see McLennan and McPhail 1989a,b, Milinski and Bakker 1990, Zuk et al. 1990c, Hill 1990, 1991, 1992, and Hill and Montgomerie 1994 for additional examples). An initial suggestion was that the expression of carotenoids in guppies could be an indicator of foraging ability (Endler 1980, 1983; Kodric-Brown 1985). Females choosing males with more carotenoids would be favored in that their offspring could inherit good genes for better than average foraging ability from those males. The condition dependence is that the expression of carotenoids in male color patterns depends on his nutritional state (= condition), which in turn depends on his foraging ability.

The obvious way to test condition dependence of carotenoids is to manipulate the nutritional status of individuals experimentally and to look at the effect on color patterns. Kodric-Brown (1989) tested specifically for an effect of carotenoid intake on the expression of carotenoid spots in guppy color patterns. In order to eliminate any additional nutritional effects, Kodric-Brown compared the effects of two diets, one with and one without carotenoids. Both diets were free of any natural carotenoids, but one diet was supplemented with an artificial carotenoid with no nutritive value. As might be expected, the density (or brightness) of orange pigment differed between diets. Fish raised on the diet containing carotenoids had significantly brighter orange spots than did fish raised on the carotenoid-free diet (fig. 6.1). Thus, expression of carotenoid spots was dependent on dietary intake of carotenoids. Kodric-Brown then showed that females preferred males with brighter carotenoid spots. Females showed a significant preference for the brighter (high-carotenoid) males in dichotomous choice tests, and then mated significantly more often with the brighter fish when they were all allowed to interact freely in an undivided aquarium (fig. 6.2). This is evidence that mate choice is based on a condition-dependent trait.

The implication of Kodric-Brown's (1989) experiment is that the female preference for brighter orange spots could be adaptive in that it identifies males that are better able to obtain carotenoids in their diet. While carotenoids themselves may or may not have major fitness effects, the relative intake of carotenoids could be correlated with the overall quality of the diet

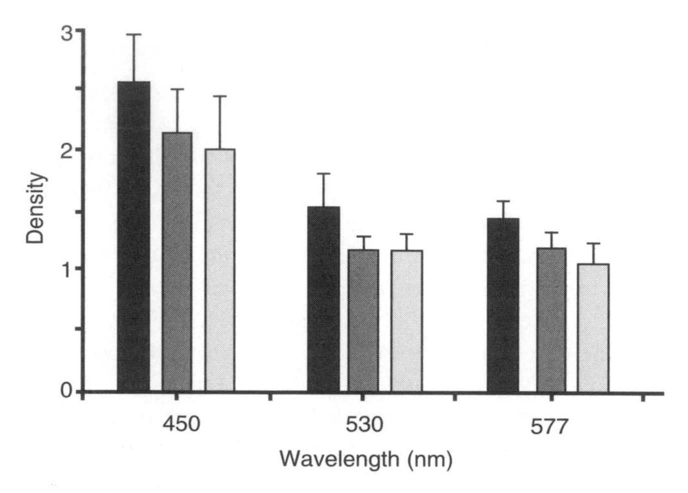

Figure 6.1 Effects of dietary carotenoids on the brightness of orange coloration. Brightness spots were measured with a densitometer at three different wavelengths using a narrow bandpass filter. The 450 nm filter corresponds to yellow; 577 nm corresponds to red. Dark bars—males on carotenoid diet; medium bars—males on non-carotenoid diet; light bars—body background color of males on non-carotenoid diet. (Adapted from Kodric-Brown 1989.) Error bars indicate one standard error.

and hence nutritional status, and the ability to absorb and retain carotenoids may also be indicative of health. Genetic variation in foraging ability could result in variation in expression of orange in color patterns similar to that produced by Kodric-Brown's manipulation of carotenoid intake.

Other experiments suggest that variation in carotenoids in color patterns could reflect more general differences in the condition of individuals in addition to differences in foraging ability. By manipulating a parasitic infection in male guppies, Houde and Torio (1992) were able to produce color differences similar to those in Kodric-Brown's (1989) study, and with similar effects on female choice. In a design involving pairs of brothers, males that had been infected with *Gyrodactylus* parasites lost some of the brightness of their orange spots relative to their uninfected brothers. As in Kodric-Brown's (1989) experiment, females preferred the brighter males over the duller males in dichotomous-choice tests, even when the fish had been cured of the infection with medication (fig. 6.3). This experiment shows that orange coloration can reflect a male's recent history of parasitic infection with effects on female choice. This kind of condition dependence is a critical assumption of the Hamilton and Zuk (1982) model, in which the evolution of female choice results from parasite-resistant genes inherited by offspring. In the guppies, females that discriminate against parasitized males might also benefit directly if there is a risk of parasite transmission in mating with an infected male (see discussion of direct benefits in chapter 5).

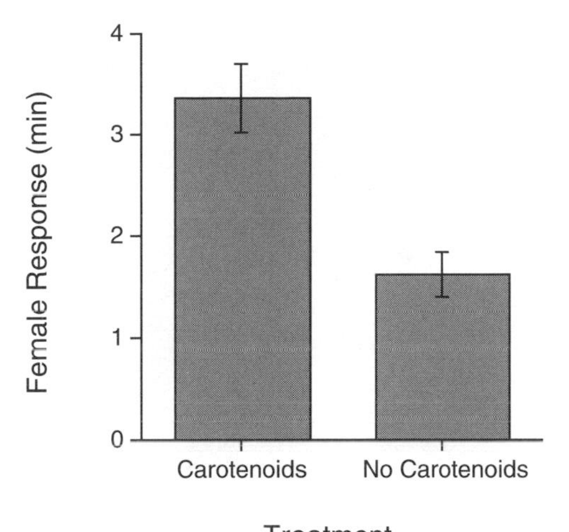

Figure 6.2 Effect of diet treatment on female preference. Bars indicate time spent in association with males in carotenoid and non-carotenoid diet treatments, respectively. (From data in Kodric-Brown 1989.) Error bars indicate one standard error.

Nicoletto (1991) has further substantiated the link between condition and expression of color patterns in guppies. He used a measure of sustained swimming performance as an index of condition and found this to be correlated with variation in the brightness of orange spots. In a subsequent study (Nicoletto 1993), orange area (and also display rate and a measure of ornament complexity) proved to be correlated with swimming performance of males as well as with the sexual responsiveness of females. Thus both correlational and manipulative experiments indicate condition dependence of the brightness of orange spots in guppy color patterns. These studies suggest that a variety of environmental stresses can affect condition and the expression of carotenoid colors. If there is heritable genetic variation in how guppies respond to those stresses, then carotenoid colors could provide cues to good genes that can be inherited by offspring.

There is extensive evidence for similar links between condition, including the effects of nutrition and parasites, on the expression of male secondary sexual characters (e.g., Møller 1988, 1990a; Zuk 1988; Borgia and Collis 1989, 1990; Clayton 1990, 1991; Zuk et al. 1990c, 1995a; Smith and Montgomerie 1991; Poulin and Vickery 1993; Bucholz 1995; see Zuk 1992, Andersson 1994, and Johnstone 1995 for reviews). Carotenoid colors are frequently implicated as condition-dependent characters that advertise male quality to females. For example, the red coloration of house finches appears to advertise nutritional status to females (Hill 1990, 1991, 1992; Hill and Montgomerie 1994). Although the main benefit of choosing red males in this species may come from differences in paternal care, heritable effects on the viability of offspring are also possible. In sticklebacks,

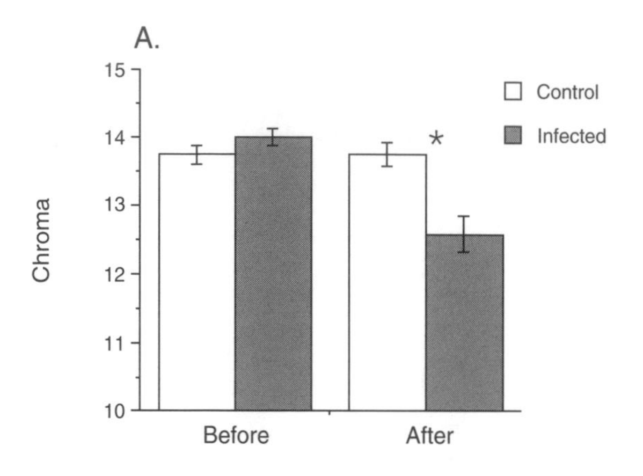

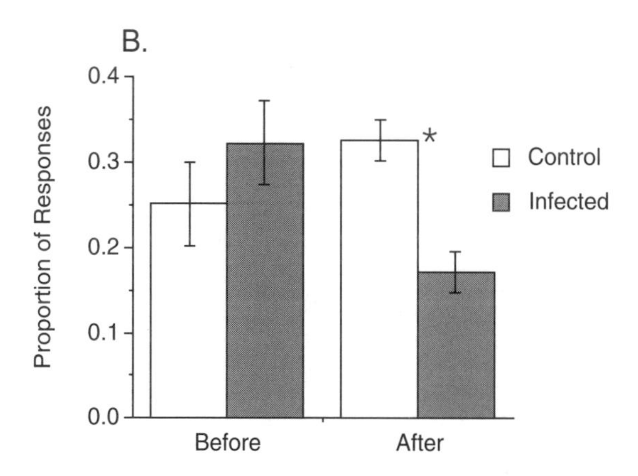

Figure 6.3 Effect of parasite infection on orange coloration and female preference. (A) Mean chroma of orange spots before and after infection or sham infection (control). (B) Responses of females to males treated as in (A), before and after treatment. (Adapted from Houde and Torio 1992.) Error bars indicate one standard error.

red throat color is related to parasite infection (Milinski and Bakker 1990; Folstad et al. 1994) and may be related to condition (Frischknecht 1993; Bakker and Mundwiler 1994). Several characters in red jungle fowl display carotenoid colors and vary depending on parasite infection (Zuk et al. 1990c).

Yet another way in which condition dependence can be assessed involves examining the degree of fluctuating asymmetry (FA) in secondary sexual traits of males (Møller 1990b; Watson and Thornhill 1994). Fluctuating asymmetry is the difference in the expression of a bilateral trait between the right and left sides of an individual, and has been suggested to be inversely correlated with heritable differences in ability to withstand environmental stresses (Møller 1990a; Parsons 1990; Møller and Pomiankowski 1993a; Watson and Thornhill 1994). Thus we can predict a negative relationship between FA and size of condition-dependent traits. This is because individuals in good condition tend both to have low FA and to produce large traits. In addition, we can predict that females might base their preferences on the symmetry of male traits. There is support for these predictions in studies of a number of species (e.g., Møller and Hoglund 1991; Møller 1992; Thornhill 1992; Swaddle and Cuthill 1994a,b). The color patterns of guppies are ideal for studies of FA. Color patterns are expressed independently on the left and right sides of the body, color pattern elements vary quantitatively between individuals, and asymmetry is readily apparent and easily measured. So far, studies of FA in guppies have not supported the predictions described above (Brooks and Caithness 1995b; Nordell 1995) and FA of color pattern elements does not seem to be a sensitive indicator of stress or genetic quality (Sheridan and Pomiankowski, unpublished data).

Effects of Mate Choice on Offspring Fitness

Despite the large body of evidence showing condition dependence of secondary sexual traits of males on which females base their mating preferences, this is not an adequate demonstration that the mating preferences have evolved through a good genes process, however. Most studies have not shown that the condition dependence of the male trait involves additive genetic variation in traits affecting the vigor of offspring.

The ultimate question needed to test good-genes models is thus whether mating preferences actually lead to an effect on the fitness (vigor or health) of offspring. Demonstrating condition dependence of the brightness of orange spots in guppies makes an adaptive function for the mating preference for this character seem plausible, but we need to demonstrate that the differences in color pattern really do reflect heritable good genes. More specifically, it is necessary to demonstrate a genetic correlation between variation in the male character and some measure of fitness of offspring (Heisler 1984, 1985; Boake 1985, 1986). Such evidence has been obtained in only a few cases (Møller 1990a, 1992; Norris 1990a,b, 1993; Thornhill and Sauer 1992; Moore 1994; Reynolds and Gross 1992).

Boake (1985, 1986) outlined an experimental method to demonstrate an adaptive effect of mate choice in which the preferences of females for a set of males are scored, the males are bred, and fitness characters of their offspring are measured. A positive correlation between the attractiveness of males to females and the relative fitness of their offspring would be evidence for a selective advantage for mate choice. Alternatively, offspring characteristics can be compared between two experimental treatment groups: one in which females are allowed a choice of mates, and one in which females are randomly assigned mates with no choice. The latter experiment has the drawback that it is not always possible to eliminate the influence of male-male competition.

These sound like simple experiments, easily carried out with guppies, but there are difficulties in methodology and interpretation of this kind of experiment (e.g., Mitchell 1990). First, traits related to fitness tend to have low heritability (Falconer 1989), so very large sample sizes are needed to measure genetic correlations between male attractiveness and offspring traits with a reasonable degree of statistical power. Second, it is not clear what measures of offspring fitness should be used. While the growth rates, age at maturity, fecundity, and other life history characters of offspring are likely to be related to overall fitness, other traits could always prove to be critical. Also, these characters measured in the laboratory may not reflect truly important differences that arise under field conditions. In this kind of study, negative results can always be explained by a failure to measure the appropriate characters under the appropriate conditions.

In the guppy system, traits such as ability to resist or cope with parasites and disease, ability to withstand periods of low food availability, or ability to escape from predators could be involved in mate choice for good genes, but are difficult to measure. In addition, differences in the more straightforward life history characters, or in survival itself, may not become apparent except under field conditions under which fish are stressed by low food, parasites, pathogens, or predators. Ideally, the fitness traits of offspring should be measured under field conditions, although this could be very difficult in practice.

Despite all these potential difficulties in performing and interpreting the good-genes experiment proposed by Boake (1985), a laboratory study by Reynolds and Gross (1992) has produced some tantalizing, though incomplete, results. Rather than basing the experiment on variation in overall attractiveness of a group of males, Reynolds focused on one particular choice criterion and its heritable effects on offspring characters. He found a preference for body size (total length) in one Trinidad population and asked whether variation in body size was correlated with the performance of offspring. This kind of experiment could be based on any character of males used as a cue in mate choice, brightness, or area of orange spots, for example.

Reynolds's finding was that offspring of larger males had higher growth rates relative to offspring of smaller males, and female offspring of large males had greater reproductive output (weight of the first two litters) (fig. 6.4). The actual attractiveness of the males, measured in terms of the sexual

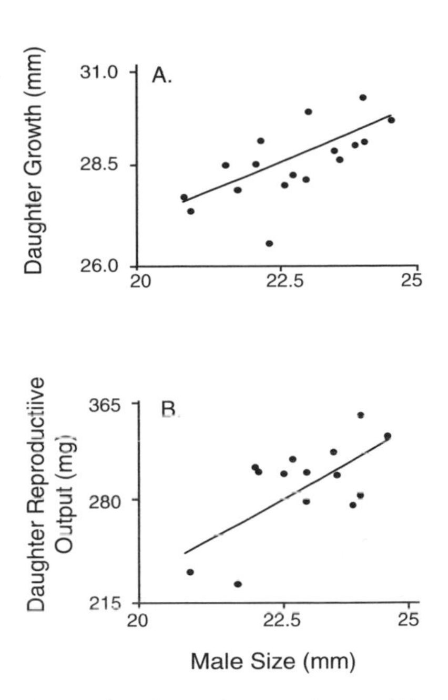

Figure 6.4 Relationship of size of males to the growth rates (A) and reproductive output (B) of their daughters. (Adapted from Reynolds and Gross 1992.)

responsiveness of females, was significantly correlated only with the growth rate of daughters, although the trends for the growth rates of sons and the reproductive output of daughters were still positive. Larger sample sizes might have shown a direct relationship between male attractiveness and offspring characters. Reynolds did not detect any preference for color pattern characters or any correlation of color pattern with offspring traits in this study.

Reynolds's results are clearly consistent with the hypothesis of adaptive mate choice. Growth rate and reproductive output are both characters likely to be closely related to fitness. On the other hand, the fact that the female choice character examined in this study was body size rather than a strictly ornamental character such as color pattern confuses the conclusions somewhat. Body size of males is an important life history character as well as a cue for mate choice in this study, and genetic correlations among size, growth rate, and fecundity are not surprising. One way of getting larger is to grow faster, and the correlation of body size and fecundity in fishes is well known. Reynolds's conclusion that a good genes process contributes to the evolution of mate choice for body size needs to be further substantiated by showing that large body size does indeed enhance the fitness of

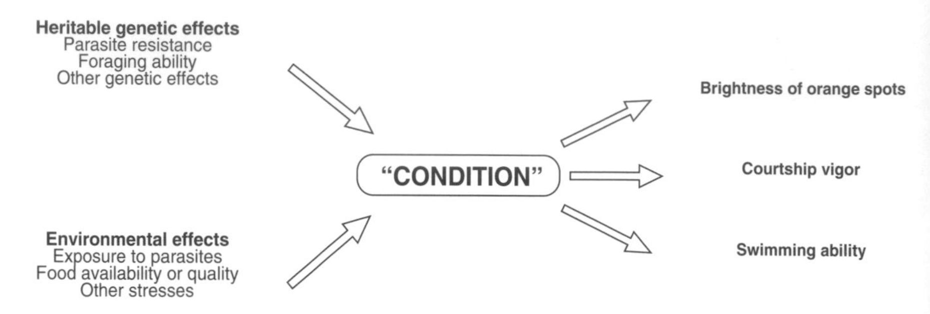

Figure 6.5 Mechanisms for condition dependence of the brightness of orange spots.

offspring. If body size were subject to stabilizing selection, then a preference for large body size probably could not be considered adaptive as suggested by Reynolds.

The good-genes hypothesis can also be tested by comparing characteristics among offspring of preferred and nonpreferred males, or among offspring of females allowed to choose their mates and females given no choice of mates. This approach has been used by Nicoletto (1995) and by Gong (1995), both of whom failed to find a difference in offspring characteristics as a result of mate choice. This lack of significant differences, especially for fitness-related characters, might have been due to small sample sizes and lack of statistical power. There is also some question as to whether females given the opportunity to choose their mates under laboratory conditions really do mate with the males they prefer in behavioral tests. In a similar experiment (see chapter 3), I found no difference in mean orange area in offspring of Paria females that were or were not allowed to choose their mates (Houde, unpublished data). Given the high heritability of orange area, this suggests that the females that could choose their mates were not actually mating with the more-orange males that they appear to prefer. If behavioral preferences as measured in the laboratory do not actually lead to differences in the mating success of males, then there would be no potential for mate choice to have heritable effects on the fitness of offspring as predicted by good-genes models.

Another strategy for testing the good-genes model is to formulate and test a specific hypothesis as to exactly how female choice for a particular character in males should have heritable effects on offspring. This is in contrast to attempting to measure all possible fitness characters of offspring as in the Boake experiment. The demonstration of heritable effects of male body size by Reynolds and Gross (1992) is an example of this approach to testing specific good-genes mechanisms. In this case, though, we need to know more about the effect of the life history traits on overall fitness. The condition-dependence experiments outlined above also suggest several specific hypotheses that could be tested. Preferences for orange coloration could be favored through heritable effects on foraging ability, parasite resistance, or possibly swimming ability. Showing that males with brighter orange spots had offspring that performed better in any of these contexts would be compelling evidence for a specific good-genes mechanism. However, it is likely that the condition dependence of orange spot coloration is mediated through more than just one mechanism (fig. 6.5). For example, evidence that the brightness of orange spots is correlated with some measure of resistance or ability to cope with *Gyrodactylus* would not rule out the possibility that another mechanism, such as the ability to survive periods of low food availability, could also play a role.

6.5 Discriminating between Good Genes and Fisherian Models

Despite debate about their relative merits, it is now clear that Fisherian and good genes models are not mutually exclusive but are most likely to operate together (e.g., Bradbury and Andersson 1987; Kirkpatrick and Ryan 1991). The question now is to what extent each process has contributed to the evolution of preferences and male characters in real systems, and to what extent each contributes to the maintenance or current "function" of preferences. Empirically, the relative importance of good genes and Fisherian processes remains largely undetermined (but see Gilburn and Day 1994), mostly because the predictions of the two classes of models are similar or identical.

Distinguishing between a good-genes and a Fisherian process for the evolution of a particular mating preference comes down to determining whether or not the preferred trait of males is correlated with heritable fitness effects ("good genes") and the degree to which the trait shows heritable variation that is independent of putative good-genes characters. Characters that are correlated with male quality could be used as cues for identifying good genes, but preferences for characters that have heritable variation not correlated with quality could be more likely to evolve through a Fisherian process alone (Grafen 1990b). This sounds like a simple distinction, but it is difficult, in practice, to rule out the possibility that a male trait indicates good genes. Condition-dependent male traits that are used as cues in mate choice do show relatively low heritability in red jungle fowl (Johnson et al. 1993) and house finches (Hill 1993), leading these authors to conclude that mate choice in these species has evolved through a good-genes mechanism.

Nicoletto (1991) attempted to assess the relative importance of good genes versus Fisherian mechanisms in guppies by measuring the extent to which male characters were correlated with a measure of quality. Nicoletto found that the density of carotenoid pigments in guppy color patterns is correlated with a measure of male vigor (swimming performance) but that variation in tail shape and dorsal fin length were not related to swimming performance. This suggests that there could be a good-genes component to preferences for carotenoid density, but that tail and dorsal fin shape would be more likely to be Fisherian characters. Nicoletto's conclusion may well be correct, although swimming performance alone is unlikely to be an adequate assay for male quality.

Area versus Brightness of Orange Spots

Similarly, the area and brightness (= carotenoid density) of orange spots are two characters that may differ in functional significance in mate choice. Much of my own work on mate choice in guppies has focused on the relative area rather than the brightness of orange spots. Unlike brightness, the area of orange is largely determined genetically (Houde 1992), independent of condition, so the Fisherian process could be relatively important. Nicoletto (1993), however, found a correlation of orange area with swimming speed, suggesting that there may be some condition dependence. On the other hand, the very high heritability estimates for this character leave relatively little scope for additional variation related to condition that is implicit in most current good-genes models. House finches appear to be similar to guppies in that there are clear genetic differences in the size of red patches among populations, but variation in the brightness of red coloration within and among populations seems to be phenotypically plastic and dependent on diet (Hill 1993).

The brightness, as opposed to the area, of orange spots in guppies may also be mainly condition dependent, so preferences for brightness could have evolved via a good-genes process. However, to measure male quality in enough ways to obtain an accurate estimate of how much of the variation in either orange area or orange brightness is or is not related to fitness will be a very difficult undertaking indeed! It would also be of interest to know how the two preference criteria, area and brightness of orange spots, interact: Do females discriminate on the basis of brightness in the same way for males that have a small area of orange as they do for males that have a large area of orange?

Although the available evidence suggests that orange brightness is more likely to be condition dependent than is orange area, the Fisherian mechanism may not be the only explanation for female choice based on orange area. Another possibility is that orange area acts as a signal amplifier (Hasson 1990) or a revealing handicap (Maynard Smith 1991) that enhances the ability of females to discriminate based on variation in the brightness of orange spots. It may be easier for females to assess brightness of large orange spots than of small orange spots, or males with larger orange spots may show a greater correlation between brightness and condition. Both of these possibilities should be testable experimentally. In either case, a preference for orange area might be considered adaptive in that it enhances the advantages of the preference for orange brightness. We can view the possibility that black spots have evolved as signal amplifiers for orange spots (see chapter 3 for details) in a similar way. Once the area of orange or black spots became involved in mate choice through this kind of process, specific preferences for these traits could evolve further through a Fisherian process.

A final, simpler alternative is that the preferences for orange brightness and orange area are manifestations of the same sensory response to orange reflectance. An increase in either area or brightness results in a greater number of orange wavelength photons striking the retina and thus might affect the female's response in the same way. This could lead to sexual selection for orange area even if the female preference initially evolved as a result of the condition dependence of orange brightness.

6.6 Variation in Mating Preferences among Populations

The guppy system provides a unique opportunity to observe geographic variation in the mating preferences of females. From a theoretical perspective this is important because the ability to detect and measure genetic variation in patterns of female choice gives us a window into the joint evolutionary dynamics of mating preferences and male characters on which preferences are based. Variation among populations allows tests of the prediction, common to both good genes and Fisherian models, that female preferences and male traits should covary within and between populations (but see Houde 1993 for discussion of limitations of between-population comparisons). We now have evidence for correlated variation in female preferences and male traits between populations of species such as guppies (see below) and cricket frogs *Acris crepitans* (Ryan and Wilczynski 1988; Ryan et al. 1992; see Bakker and Pomiankowski 1995 for other examples). Evidence for genetic variation in mating preferences within populations has been more difficult to obtain (Pomiankowski and Sheridan 1994a,b; Breden et al. 1994; see examples in Bakker and Pomiankowski 1995), with the greatest success coming from selection experiments (Breden and Hornaday 1994; Houde 1994—see below; Gilburn et al. 1993; Wilkinson and Reillo 1994).

From a practical point of view, geographic variation in preferences means that the mechanisms of sexual selection and the evolution of female choice that are documented in a single study of one population are not generalizable to all populations of guppies (Endler and Houde 1995). This may be a reason for some of the inconsistent results between studies of sexual selection in guppies. On the other hand, an understanding of how and why sexual selection processes can differ between guppy populations can lead to a broader understanding of the relative importance of different evolutionary mechanisms and their interplay across species.

In her initial experiment demonstrating female choice based on color pattern characters, Kodric-Brown (1985) found a remarkable degree of unanimity in the preferences of individual female guppies from two different laboratory strains and from a domesticated wild-type stock. These females all showed consistent preferences based on the carotenoid spots of test males. Given that there are plausible reasons why female preferences for carotenoids in the color pattern could be adaptive, Kodric-Brown argued that the similarity of preference between different strains is likely to reflect similar patterns of good-genes selection acting on female choice. Under a purely Fisherian model, female preferences are likely to be based on arbitrary traits that are not necessarily similar between populations. This argument is reasonable, but Kodric-Brown's data were not sufficient to close the case. The laboratory and domestic strains that Kodric-Brown studied might not have been representative of natural variation between populations; the preference data consisted of relative rankings of males and would not have detected differences in degree of preference; and there was essentially no direct information about condition dependence of carotenoid colors in guppies at the time of the study. Subsequent studies have supported the contention that carotenoid colors can be condition dependent (see above), and have shown that preferences for carotenoids are common but not universal among guppy populations (Houde and Endler 1990; Endler and Houde 1995).

Several studies have compared mating preferences between guppy populations that differ in color pattern (Breden and Stoner 1987; Stoner and Breden 1988; Houde 1988a; Houde and Endler 1990; Endler and Houde 1995). I was initially intrigued by the striking differences in expression of carotenoid pigment between populations in my lab. My early experiments demonstrating a preference for males with greater relative area of orange in their color patterns (Houde 1987) were all performed on descendants of guppies originally collected in the Paria River of Trinidad. Color patterns in the Paria population are typically dominated by especially large areas of orange, although there is considerable variation. A number of other populations have much less orange in their color patterns and more of other colors. So the natural empirical question was whether females from populations with less orange would show similar preference for orange.

I began (Houde 1988a) by comparing the preferences of females from the Paria population to preferences of females from a high predation locality in the Aripo River. Aripo males have relatively little orange in their color patterns (fig. 6.6). To compare preferences, I tested a group of males (all from the same population) with, say, Paria females on one day, and then with Aripo females on another day. This allows male-by-male com-

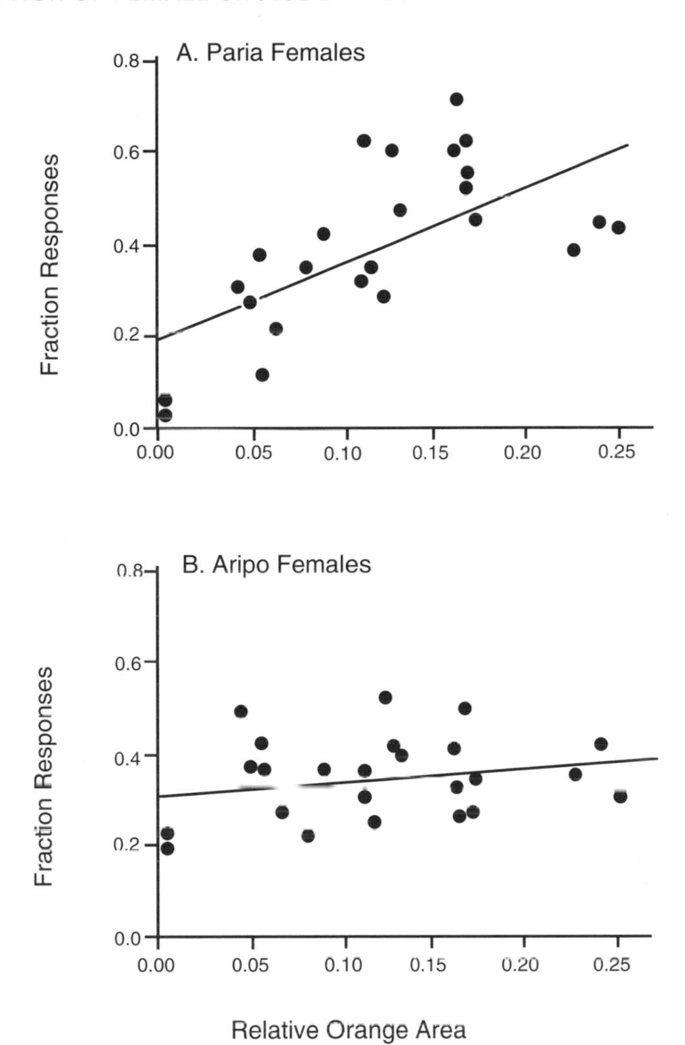

Figure 6.6 Comparison of mating preferences for orange in Aripo and Paria females. (Adapted from Houde 1988a.)

parison of female preferences. There was a clear difference in preference, measured by variation in responsiveness of females, between the two populations. As before, Paria females were most responsive to males with relatively large amounts of orange and discriminated against the less orange males. But Aripo females were equally responsive to all males regardless of orange area. A significant difference in slopes of regression of female responsiveness on orange area confirmed the difference in preference (fig. 6.6). Because the females had all been reared under similar conditions in the laboratory, this can be interpreted as a genetic difference and evidence of evolutionary divergence in preference. Interestingly, further analysis revealed that not only did Aripo females show no preference for orange, but they also discriminated very little among particular males. There was much greater variation in the apparent attractiveness of Paria males than of Aripo males. This points out that, had I studied Aripo guppies initially, I might have obtained results much more consistent with those of Farr (1980b) or Reynolds and Gross (1992), i.e., no evidence for mating preferences for particular color patterns or color pattern characteristics.

Breden and Stoner (1987) and Stoner and Breden (1988) also documented differences in female preferences between populations. Both studies compared preferences of females from high- and low-predation populations that differed in male color pattern and were motivated by the prediction that variation in preferences should be correlated with variation in color patterns. Consistent with this prediction, Stoner and Breden (1988) found that low-predation females tended to favor the "brighter" of two males in dichotomous choice tests (where "brightness" is measured as the total area of colored spots). Females also preferred the more active, or responsive, of two males. Females from high-predation populations, in which conspicuous color patterns and active courtship entail a greater risk, showed no preference for bright or actively courting males. The difference in behavior of the females was statistically significant and thus indicates a genetic difference in preference. Breden and Stoner (1987) obtained similar results when they compared the response of high and low predation females given a choice between a moving, "bright" model guppy and a stationary, "dull" model. In this case, high-predation females preferred the "bright" model and low-predation females preferred the "dull" model. Although the design and interpretation of this experiment have been debated (Endler 1988; Breden 1988; Houde 1988c; Breden and Stoner 1987), the results are consistent with the predictions from theory and with experiments using different design and populations (Houde 1988a; Stoner and Breden 1988; Houde and Endler 1990; Endler and Houde 1995).

The main conclusion drawn from these comparative studies is that although guppy color patterns may reflect a balance between mating advantage (sexual selection) and risk of predation (natural selection), the evolutionary outcome is not the same in all populations and we cannot think of mating preferences typologically. Instead, both mating preferences and color patterns seem to be evolutionarily labile. But do these two traits evolve independently of each other?

My initial results, and those of Breden and Stoner, also suggested that there was a correspondence between variation in preferences and variation in color patterns in general. This motivated a much larger comparison of

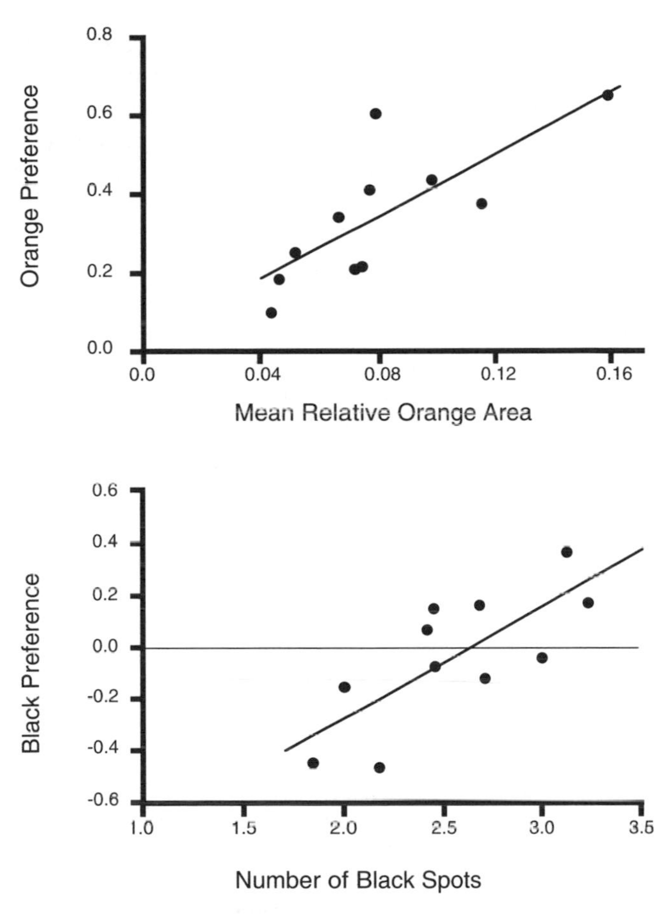

Figure 6.7 Relationship of female preferences and mean male characteristics in a comparison of eleven guppy populations. (Adapted from Endler and Houde 1995.)

patterns of female choice across populations (Houde and Endler 1990; Endler and Houde 1995). John Endler and I measured female preferences and scored male color patterns in an additional ten populations, and we have looked at effects of other color pattern and morphological variables besides orange area.

Our results (Houde and Endler 1990; Endler and Houde 1995) confirmed that there is extensive variation in preferences between guppy populations. Furthermore, variation in preference seems to parallel variation in the orange area (fig. 6.7), confirming the hunch provided by the initial comparison of Paria and Aripo preferences. The number of black spots also parallels the degree of preference for black spots across populations. Preferences for orange occur in several populations (fig. 6.7; Endler and Houde 1995), suggesting that a widespread but not universal process may be at work. Preferences based on other color pattern characters and on size variables have been documented only in isolated cases, however. We may have failed to detect preferences in some cases due to small sample sizes and large error inherent in measuring preferences, so more intensive studies of particular populations are needed.

Differences in mating preference between populations are consistent with the Fisherian model only in that there appears to be no single optimal preference. Differences in preferences could occur under other sexual selection models as well (see chapter 5), so variation alone is not sufficient evidence to support the Fisherian model (Houde 1993). The fact that variation in preferences for orange area parallels variation in orange area itself (Houde 1988a; Houde and Endler 1990; Endler and Houde 1995) is consistent with both Fisherian and good-genes models, and suggests a specific way in which the Fisherian mechanism may operate in the guppy system. In guppies, the color patterns of males are clearly subject to direct selection by predators (see chapter 1), so the Fisherian model predicts an indirect response in female mating preferences. This should lead to correlated variation in female mating preferences and male color patterns between populations that experience different regimes of predation. This argument is compelling but does not rule out the possibility that direct selection or a good-genes mechanism may be responsible for the origin or subsequent modification of female preferences. Virtually any model for the evolution of female mating preferences predicts correlated variation in preferences and male characters on which preferences are based (Houde 1993). In addition to the Fisherian model, in which preferences change in response to changes in the male character, any mechanism that leads to change in preferences should result in corresponding changes in the male character. Even though the data on parallel variation in preferences and male color patterns do not discriminate among sexual selection models, they do have important implications. Both traits vary among populations, but they vary nonindependently, in parallel, showing that their joint evolution is constrained by the effects of sexual selection.

Mismatch between Preferences and Male Traits

Despite the theoretical expectation that secondary sexual characters of males should evolve to match the preferences of females and vice versa, there are a few examples in which females have preferences for traits that are not strongly expressed or not expressed at all by males in their species. These include the preferences of female platyfishes (e.g., *Xiphoporus maculatus* and *X. variatus*) and related species (e.g., *Priapella olmecae*), in which males have unelaborated tails, for swordlike extensions of the tail seen typically in swordtails (e.g., *Xiphophorus helleri*; Basolo 1990a,b, 1995a,b). In another example, *X. pygmaeus* females prefer *X. nigrensis* males over conspecific males. This may be the result of a preference common to females of both species for large body size and courtship displays, even though *X. pygmaeus* males do not court or show large body sizes comparable to *X. nigrensis* (Ryan and Wagner 1987). Finally, in the Neotropical frog *Physalaemus pustulosus*, males produce a two-part call in which the second part, the "chuck," is preferred by females. In other related *Physalaemus* species, however, males do not produce chucks, but females demonstrate a preference for calls that include chucks in laboratory tests (Ryan and Rand 1990; Ryan et al. 1990b). All of these examples include species in which females have a preference for a male characteristic not exhibited by conspecific males, although it is exhibited by males in related species. The mismatch between female mating preferences and male characteristics in these examples is in contrast to the correlated pattern of preferences and color patterns seen in guppies. Also, in all of these examples, phylogenetic evidence suggests that the female preference existed prior to the evolution of the preferred male characteristic. This supports the idea that preexisting biases in the sensory systems of females can result in mating preferences that lead to the evolution of male traits in the process of "sensory exploitation" (Basolo 1990a; Ryan 1990; Ryan et al. 1990b).

There may be exceptions to the general correlated pattern of expression of color patterns and preferences in guppies too. In figure 6.7, at least one population seems to have relatively low expression of orange in males for the level of preference shown by females. This data set (Endler and Houde 1995) was not sufficient for individual comparisons, but a subsequent comparison of another population, from a high-predation locality on the lower Yarra River, confirmed that color patterns do not always match female preferences in guppy populations. In the Yarra population, we found a strong preference for orange area, equal to that seen in the Paria population, but significantly less expression of orange coloration than in the Paria population (Houde and Hankes, in press). This suggests that the behavioral preferences revealed by laboratory experiments do not always lead to the evolution of corresponding male traits, in this case large areas of orange in the color pattern.

A number of possible constraints on the evolution and expression of female preferences and on their effects on the evolution of male traits could lead to this apparent mismatch. If the predation regime has changed recently or the Yarra population is the result of a recent introduction, we might expect rapid change in highly heritable color patterns but slower response in mating preferences to produce a temporary mismatch. Life history traits in the lower Yarra River are consistent with the present regime of high predation (Reznick and Bryga 1996; Reznick et al. 1996). Thus there should have been ample time for evolution of preferences as well as color patterns and life history traits. On the other hand, a high degree of divergence from other guppy populations and low allozyme variability in guppies from the upper Yarra River supports the possibility that the Yarra River could have been recently colonized by a small founder population (Carvalho et al. 1991, 1996; Magurran et al. 1995). Finally, it is possible that females do not or cannot translate the mating preferences shown in the lab to mate choices in the natural population in the Yarra River. For example, females may be modifying their mating preferences in the presence of the abundant large predatory fishes in this river (Godin and Briggs 1996; see chapter 5). Another possibility is that aspects of the social environment such as the high schooling tendency in Yarra guppies (pers. obs., see also Magurran et al. 1995) or sneak copulation behavior of males might constrain the ability of females to mate with the males they prefer. The reasons why male traits sometimes do not correspond to female preferences need to be investigated further in guppies and other species.

A similar situation exists in house finches and sticklebacks and suggests an additional explanation for the mismatch in Yarra guppies. In most stickleback populations, males have red throat patches and females base their preference on the red patches (see chapter 3). In a number of populations in western North America, males have black rather than red throat patches (Reimchen 1989). Females from both populations prefer males from red populations over those from black populations, although females from black populations may express this preference to a lesser degree (McPhail 1969). Tests in which females compare males from different populations might confound coloration differences with differences in other traits, so McKinnon (1995) used video-editing techniques to present males that differed only in throat color to females. He found that females from both populations preferred males with red or black throat patches over males with no throat patch. This appears to be a situation, like that of the Yarra guppies, in which females in two populations have similar preferences but males differ in the preferred sexual trait. In sticklebacks, one possibility is that red and black throat patches are equivalent in how they are perceived by females, especially if water color alters the appearance of males to females (Reimchen 1989; McKinnon 1995; see discussion of sensory drive in chapter 5). Another possibility, which might apply to guppies, is that dietary carotenoids are too limited in some populations to produce sufficiently red throat patches (Reimchen 1989), and black throat patches have evolved in these populations as the next most attractive sexual ornament. A similar argument may apply to house finches.

Although most of the variation in red coloration among populations of house finches appears to be related to availability of dietary carotenoids, the difference in the size of red patches between two subspecies, *C. m.*

frontalis (widespread in North America) and *C. m. griscomi* (a Mexican population) appears to be genetic (Hill 1993). Females from both subspecies prefer males with large, bright patches of red coloration (Hill 1994). The size of red patches in *C. m. griscomi* thus does not correspond to the preference of females. The most likely explanation, Hill (1994) argues, is that *griscomi* males have evolved under conditions with chronically low availability of carotenoids in the diet and are unable to produce bright red coloration in large patches. Given that females base their preference on the brightness as well as on the size of red patches, Hill (1994) argues that sexual selection favors *griscomi* males that express smaller but brighter red patches. This is similar to Reimchen's (1989) suggestion for sticklebacks. An analogous argument could be made for Yarra guppies and would be amenable to experimental testing. However, generality of the argument that a male trait could evolve to become mismatched with the female preference in such a scenario needs further theoretical exploration.

Given that the correspondence between female preferences and male color patterns is not universal in guppies (although it seems to be a general trend), it is also of interest to ask if the female preference for orange coloration could represent a preexisting sensory bias present in ancestral guppies (see Basolo 1990a). At present we do not know enough about the phylogeny of guppy populations to provide a clear answer. What phylogenetic information we do have (fig. 1.7) suggests that a preference for orange may have originated more than once or may have been lost more than once, but we cannot determine whether there could have been a preexisting bias for orange coloration in ancestral guppies. We know that preferences for orange are present in Yarra guppies, Paria guppies, one population from the Caroni drainage, and several populations from the Oropuche drainage (Endler and Houde 1995); these represent the main phylogenetic divisions among guppies (Magurran et al. 1995). Preferences for orange are lacking in populations from both the Oropuche and Caroni drainages (Endler and Houde 1995). This suggests at least that mating preferences are evolutionarily labile in guppies. Frequent evolutionary changes in mating preferences are implied by Fisherian models (Iwasa and Pomiankowski 1995), but could arise by other mechanisms as well.

6.7 Genetic Correlation of Preferences and Color Patterns within Populations

We have seen that finding a correlation between female mating preferences and male color patterns demonstrates that the evolution of these two traits is interdependent but does not test specific sexual selection models. A more direct way of testing for genetic correlation between female preferences and male traits within populations, an assumption common to both good-genes and Fisherian models, is through artificial selection experiments or by testing for a correlation of sexual traits in males with mating preferences of their daughters. Two selection experiments with guppies provide somewhat ambiguous results (Breden and Hornaday 1994; Houde 1994). Similar data from other systems also support the role of correlated evolution in sexually selected systems (sticklebacks: Bakker 1993; seaweed flies: Gilburn et al. 1993; stalk-eyed flies: Wilkinson and Reillo 1994; see Bakker and Pomiankowski 1995 for a review).

In the guppy system, the obvious experiment to test for the predicted genetic correlation was to impose artificial selection on male color patterns and to see if the predicted response in female preferences occurs. In my experiment (Houde 1994), I imposed artificial selection for increased or decreased orange area on four pairs of high (increased orange) and low (decreased orange) selection lines over three to four generations. The lines were initiated using either wild-caught or offspring of wild-caught guppies from the Paria River of Trinidad. In each generation, males were selected according to orange area and females were chosen at random. Given the very high heritability of orange area (Houde 1992), the immediate and dramatic response to selection on orange area in most lines was not surprising. Fisher's model of sexual selection argues that, because of nonrandom mating, males with a given orange area in their color pattern should also carry genetic information for a corresponding level of preference inherited from their mothers. As a result, selection on orange area should lead to corresponding changes in the strength of female preferences. At each generation, I scored the preferences of low-line and high-line females and compared them as in the population comparison experiments. The selection experiment by Breden and Hornaday (1994) used a similar method, differing in population of origin, sample sizes, color pattern trait scored, and details of breeding.

In my experiment (Houde 1994), all four pairs of selection lines showed divergence in female preferences in the first two generations. In all cases, females from the lines selected for increased orange area exhibited stronger preferences relative to females from the lines selected for decreased orange area. A shift in preference as a correlated response to selection on the male character appeared to have taken place. Unexpectedly, however, the divergence in preference was no longer significant after the third generation in two of the four pairs of lines (fig. 6.8). In their artificial selection experiment, Breden and Hornaday (1994) found no evidence for divergence in female preference after five generations of selection. A possible reason for the inconsistent results after the third generation in my experiment, and for the lack of correlated response in Breden and Hornaday's (1994) experiment, could be that females were not able to exercise their preferences in the laboratory, leading to a breakdown in the genetic correlation in later

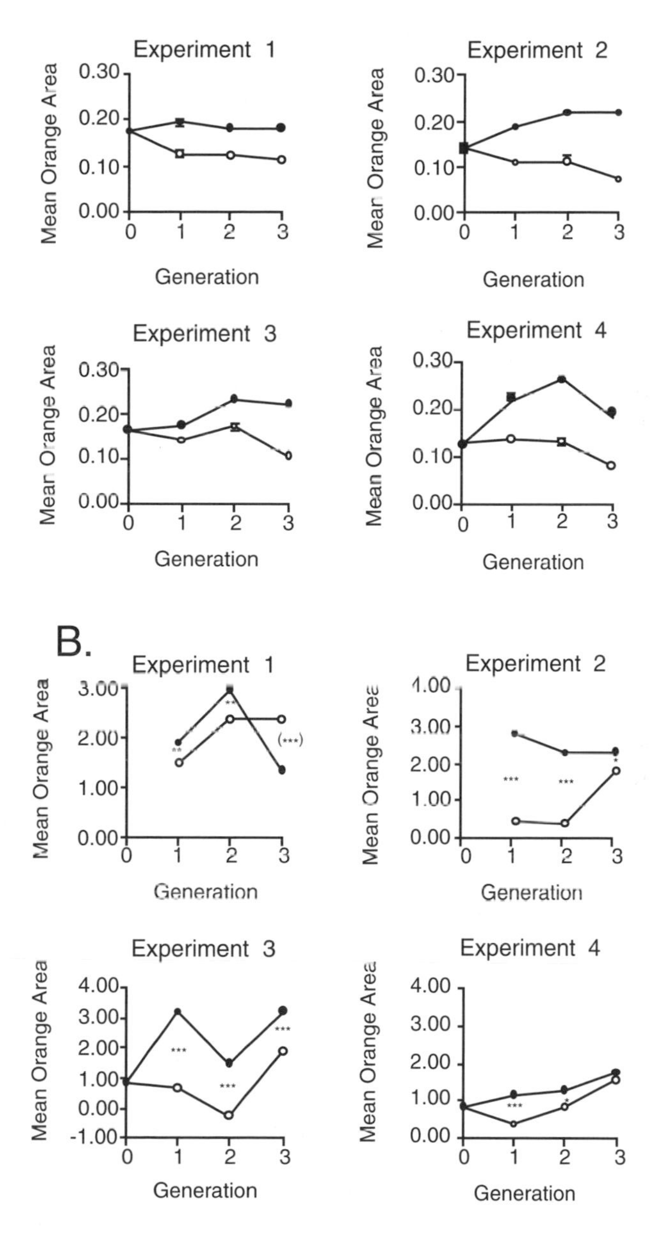

Figure 6.8 Results of artificial selection for increased and decreased orange area in male color patterns (from Houde 1994). Results are for four pairs of high- and low-selection lines (Experiments 1–4). (A) Changes in mean orange area. Differences between high and low lines are all highly significant. (B) Changes in female preference. Asterisks indicate degree of departure from expected divergence due to genetic drift alone. (See Houde 1994 for further details.)

generations. This breakdown in the genetic correlation could be especially likely in the laboratory with small effective population sizes (twenty selected males per generation in my experiment, five in Breden and Hornaday's experiment; Nichols and Butlin 1989; Gilburn and Day 1994). Alternatively, the difference in results between the two experiments could reflect real differences between populations in the variance-covariance structure of the two traits (see also discussions by Pomiankowski and Sheridan 1994a and Bakker and Pomiankowski 1995).

The results of these selection experiments, although somewhat ambiguous, do provide some support for the genetic correlation between female preferences and preferred male characters predicted by both Fisherian and good-genes models. The artificial selection regimes simulate part of the dynamics of a Fisherian process, in which the male character is subject to natural selection, and the female preference evolves as a correlated response. This suggests that the Fisherian process is plausible, but does not prove that it has actually occurred. The second part of the Fisherian process, in which changes in female preference lead to further change in the male character, has yet to be demonstrated.

6.8 Sexual Selection and Speciation

Our evidence for evolutionary lability in female mating preferences has implications beyond specific sexual selection models. The divergence of mating preferences can potentially lead to reproductive isolation between populations and eventual speciation (Lande 1981; Kirkpatrick 1982; West-Eberhard 1983; Lande and Kirkpatrick 1988; Kirkpatrick and Ryan 1991; Schluter and Price 1993; Iwasa and Pomiankowski 1995). Darwin himself recognized this as a general mechanism, suggesting that intersexual selection could even be important in the divergence of human populations (Darwin 1871, chapters 19 and 20). The specific conditions under which divergence in preferences like that seen in guppies can lead to speciation is not well understood, however.

Premating isolation between populations can arise if female mating preferences and corresponding male traits become so different that males from one population are no longer attractive to females from the other population, and vice versa. Guppies may represent an early stage in this process. Female mating preferences have diverged among populations, as have male color patterns and other traits, and this may lead to assortative mating when individuals from different populations meet. For example, females from a population with a strong preference based on orange coloration might discriminate against most males from a population in which some other color pattern element is strongly expressed but orange is not.

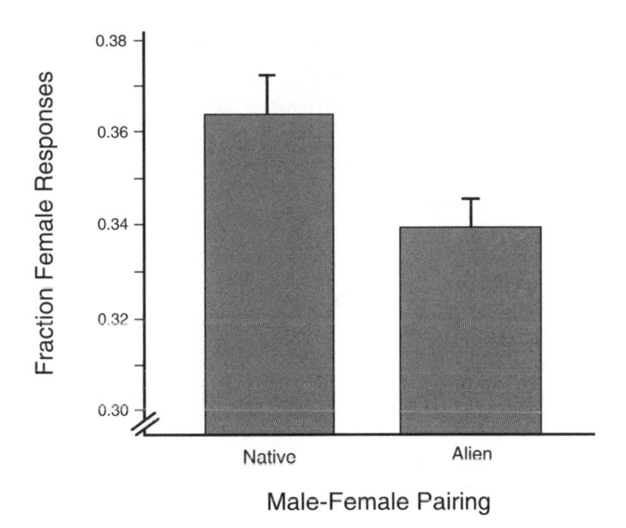

Figure 6.9 Responsiveness of females to the displays of males from their own ("native") population and to males from other ("alien") populations. (Adapted from Endler and Houde 1995.) Error bars indicate one standard error.

In laboratory experiments, female guppies are slightly but significantly more responsive to native males (from their own population) than to alien males (from a different population) (Houde 1988a; Endler and Houde 1995; see fig. 6.9). In tests where native and alien males competed for matings, native males had an advantage (Luyten and Liley 1991). This pattern of assortative mating may be the result divergent mating preferences, but male discrimination might also play a role (Luyten and Liley 1991). We do not know, however, if these apparent preferences for native males are the result of divergent preferences for color patterns or divergent preferences based on courtship behavior or other morphological traits. Another possibility is that a combination of size-assortative mating and body size differences between populations could produce an apparent preference for native males (Endler, pers. comm.). Guppies from different populations mate readily and produce viable offspring (pers. obs.), although detailed studies of the success of hybrids would be interesting.

Mate choice by males and by females does seem to play a role in reproductive isolation between guppies and their closest relatives (*Poecilia picta, P. parae, and P. vivipara*), which occur sympatrically in some places (Haskins and Haskins 1949; Liley 1966). Does this interspecific mating discrimination represent a more extreme divergence of mate choice patterns than that seen among guppy populations? Further studies of the basis for discrimination among guppy populations and among these species are needed in order to address this question.

Why has the divergence of guppies not proceeded beyond a preliminary stage in speciation? There are examples of other taxa in which sexual selection may have led to full reproductive isolation and sympatric coexistence. The numerous Lake Victoria cichlid species, for example, have been diverging for only somewhat longer than guppy populations (500,000–750,000 years for cichlid species: Meyer et al. 1990; 330,000–500,000 for guppy populations: Carvalho et al. 1991; Shaw et al. 1991, 1992; Fajen and Breden 1992; Magurran et al. 1995). The Lake Victoria cichlids have diversified into about three hundred reproductively isolated, phenotypically divergent, sympatrically coexisting species, while guppy populations, although nearly as ancient, have retained their ability to interbreed, show more subtle phenotypic divergence, and thus do not show sympatric diversification. There are closely related *Poecilia* species (see chapter 1) that occur sympatrically with guppies in lowland and brackish habitats (Haskins and Haskins 1949; Liley 1966), but there are no sympatric species in the small mountain streams in which most studies of guppies have been conducted.

Another interesting comparison is between guppies and other river- and stream-dwelling fishes with conspicuous color patterns. The North American darters (Kuehne and Barbour 1983) and African killifishes in the genus *Diapteron* (Brosset and Lachaise 1995) are both groups with sympatrically coexisting species in which male color patterns are strongly differentiated, implying a role for sexual selection in speciation. Again, we can ask why speciation followed by sympatric coexistence has occurred in these groups while guppies have differentiated but have not speciated. One possibility is that the potential for speciation in guppies was limited because, at the time of initial colonization, there were relatively few unfilled niches for fishes (Endler 1983 and pers. comm.).

6.9 Summary

Much of the work on female choice in guppies was initially motivated by a research agenda that attempted to discriminate among sexual selection models (see Bradbury and Andersson 1987). Although direct selection may shape the mating preferences of female guppies in important ways, as we saw in chapter 5, indirect selection models (good genes and Fisherian) have been viewed as more likely to explain the origins and maintenance of female preferences in guppies. The view of good genes and Fisherian models as alternative explanations has been replaced by a more integrated concept in which both processes may contribute simultaneously. Results from studies of guppies undoubtedly contributed to the evolution of our theoretical understanding, and have also provided several important empirical conclusions.

One contribution has been to confirm the fundamental assumption and prediction that female mating preferences vary genetically and can undergo evolutionary change. Evidence for this comes from studies showing significant repeatability of preferences of individual females, an artificial selection experiment, and comparisons of mating preferences among populations. Despite considerable variability and apparent stochasticity in the behavior of females, the mating preferences of individual females are consistent in repeated observations and differ significantly among females. Divergence in the mating preferences of females from a population subjected to artificial selection confirmed that differences in preferences have a genetic basis and that preferences have the potential to evolve in response to direct or indirect selection. Finally, differences in female preferences among populations demonstrate the that this trait does indeed evolve in nature.

A number of studies have provided evidence in support of the good-genes mechanism for the evolution of female mating preferences in guppies. Most of the evidence comes from studies showing that the brightness and area of orange spots are condition dependent, that they are related to aspects of health and vigor. The expression of orange colors by male guppies is affected by parasite infection and nutrition and is correlated with swimming ability. As yet there is no direct evidence that the condition dependence of orange coloration in guppies translates into fitness effects that can be inherited by offspring, nor has such an effect been demonstrated for carotenoid colors in any other species. Further work on the possible association of carotenoid colors in males with fitness characters in offspring is therefore needed, and guppies are a promising system for such a study. Mate choice based on male body size in one guppy population does affect the fecundity of daughters, however, and may have evolved through a good genes mechanism.

Finally, guppies have provided support for the idea that the evolution of female mating preferences and male sexual traits are interdependent. Consistent with theoretical arguments and predictions, these two traits show a tendency to evolve in parallel across populations and to covary genetically in at least some populations. The parallel variation in mating preferences and color patterns is predicted by several sexual selection models but does not distinguish among the models. The correlated pattern of variation in guppies does provide counterpoint to cases in which female preferences and male traits do not coevolve (e.g., *Xiphophorus* and *Physalaemus*). In some cases like guppies, sexual selection processes seem to drive the evolution of both female preferences and male traits as predicted by theory, while in other cases, these characters seem to evolve independently. In fact, even in guppies, there are populations in which color patterns do not match female preferences. Although a number of hypotheses could explain why male traits do not always match female preferences, empirical studies and further theoretical work are needed to address this question.

Female mating preferences and male traits are genetically correlated within at least some guppy populations. The existence of this genetic correlation is an essential argument of indirect selection models and is a necessary condition for the operation of good genes and Fisherian processes. In guppies, support for the correlation comes from experiments in which female mating preferences showed an indirect response to selection on male color patterns. Only a few other studies have shown support for a genetic correlation between female preferences and male traits within populations. Only one of two guppy studies found evidence for the correlation, however. This inconsistency might reflect real differences among populations in the genetic variances and covariance of the two traits. Further comparative studies would be of interest to determine whether the variance-covariance structure for female preferences and male sexual traits differs among populations.

Evolutionary changes in mating preferences like those documented in guppies have the potential to contribute to reproductive isolation and speciation. There is some suggestion of premating isolation between guppy populations, but ecological constraints probably prevent guppies from fully speciating and undergoing adaptive radiation. Guppies provide a useful model system for studying the early stages of speciation, however, and further studies are needed to understand how divergence in preferences leads to isolation among populations.

7 Summary and Prospects

7.1 What We Have Learned about Sexual Selection in Guppies

We owe a great deal to the pioneers in research on guppies, especially Caryl and Edna Haskins, and Robin Liley (see especially Haskins and Haskins 1949, 1950; Haskins et al. 1961; Liley 1966), for recognizing the potential of this species for addressing issues in evolution, behavior, and ecology. The early studies of guppies addressed questions about the role of mate choice in reproductive isolation, the evolution of geographic variation in color patterns and other characters, and the role of sexual selection in the evolution of color patterns. This monograph has focused primarily on the latter issue: sexual selection and the evolution of color patters and other secondary sexual traits. Since the seminal studies cited above, our picture of how sexual selection affects evolution in guppies has become more complete, more detailed, more complex, and perhaps more clear.

Studies of sexual selection in guppies began with a relatively simple view of the evolution of color patterns and courtship behavior of males (see fig. 7.1A). The original assumption was that the expression of color patterns and courtship behavior represented a simple balance between the evolutionary effects of natural selection (predators) and sexual selection (female choice or male-male competition). The studies discussed in the preceding chapters have revealed that the situation is considerably more complex. Not only do factors other than predators play a role, but predators and other factors are responsible for individual plasticity in behavior as well as for evolutionary divergence among populations.

I have attempted to describe the current state of understanding of sexual selection processes in guppies in figure 7.1B. This diagram contains the elements of the original view (fig. 7.1A) at its core, but shows the additional understanding we have gained about the evolutionary processes and patterns of phenotypic plasticity affecting mating preferences and secondary sexual traits. Each arrow in figure 7.1B refers to an evolutionary process leading to genetic change in populations, to plasticity in expression of

Sexual selection processes in guppies

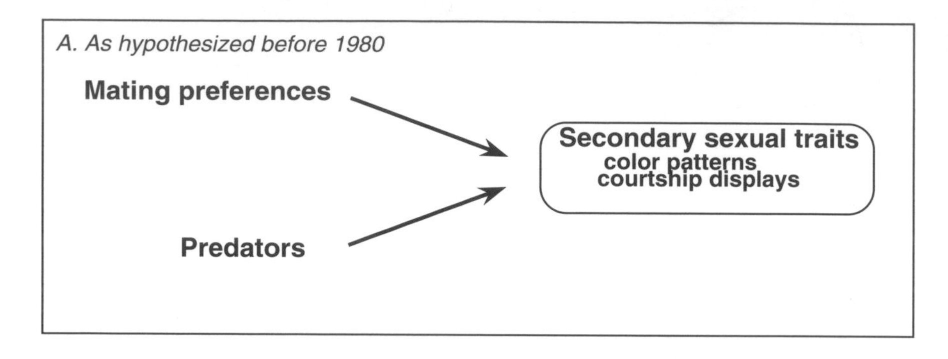

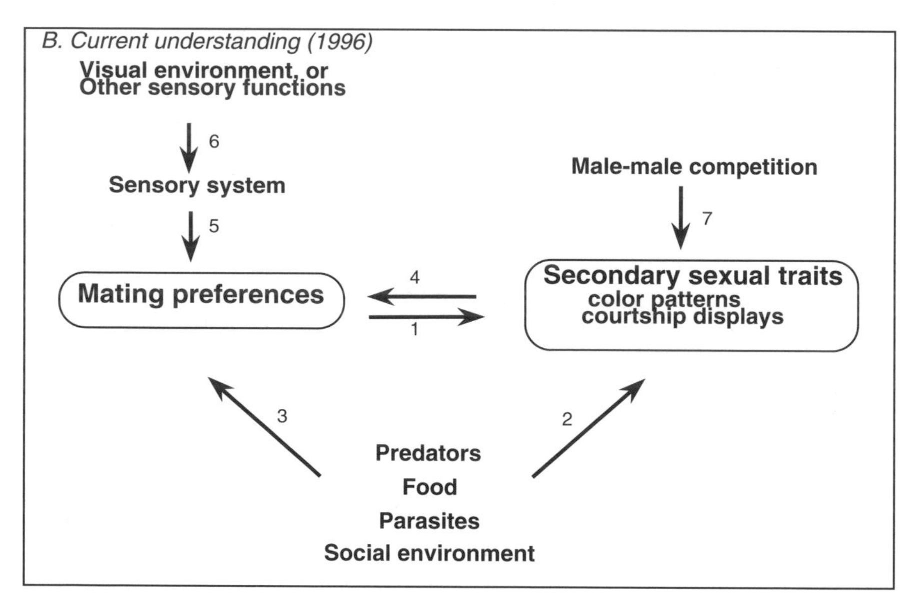

Figure 7.1 Our understanding of sexual selection processes in guppies.

traits within individuals, or both (referred to in the following text as "evolutionary effects" and "plasticity," respectively). In the discussion that follows, I have provided section references so the reader may refer back to previous chapters. The relationships depicted in figure 7.1B are not limited to the guppy system but are generally applicable to most species in which sexual selection has been studied.

Studies of sexual selection in guppies began with the assumption that sexual selection leads to the evolution of conspicuous color patterns and courtship displays (Arrow 1: evolutionary effect; fig. 7.1A and B) and that this sexual selection is opposed by natural selection (effects of predation; Arrow 2: evolutionary effect, and fig. 7.1A). Thus, variation in sexually selected characters (especially color patterns) could be attributed primarily to variation in predation regime (Arrow 2: evolutionary effect) among guppy populations.

These basic hypotheses have been confirmed in many studies. Some of the first good evidence for female mating preferences based on secondary sexual traits of males (Arrow 1: evolutionary effect) came from work on guppies (section 3.3). The inference that mating preferences lead to sexual selection with evolutionary effects on color patterns is also reasonably well supported in guppies (sections 3.2 and 3.3). There is convincing evidence for the role of predation as an agent of selection on color patterns (Arrow 2: evolutionary effect) and other traits in guppies (section 1.2), leading to geographic variation corresponding to differences in predation regime. Courtship displays of males are probably subject to the same kinds of selection as are color patterns (Arrows 1 and 2: evolutionary effects), but the evidence is not as extensive or clear-cut (section 4.2). In addition to the effects of predation regime represented by Arrow 2, genetically based geographic variation in secondary sexual traits might also reflect varying costs and benefits due to other factors. These could include differences in food availability, parasites, or social structure between populations, but these possibilities remain to be investigated in guppies.

We do know that there is adaptive plasticity within populations in the expression of color patterns and display behavior in response to these factors, (Arrow 2: plasticity), in addition to the evolutionary responses to selection. Male guppies adjust their courtship behavior in response to risks of predation, food availability, parasite infection, and differences in social structure of their group of conspecifics (sections 4.2, 4.3, and 4.4). Environmental factors such as food availability or parasites may constrain the expression of male color patterns or courtship, either directly or through their effects on the overall condition of males (Arrow 2: plasticity). For example, the quality of diet or the presence of parasites or disease can reduce the expression of carotenoid coloration and courtship display (section 6.4). This condition dependence supports good-genes models for the evolution of mating preferences (see section 6.3), in that it may lead to indirect selection on female mating preferences (Arrow 4: evolutionary effect). Males also have moment-to-moment control over the appearance of their color patterns in response to the behavior of females and possibly also to other males and predators (Arrow 2: plasticity; section 2.3).

Selection can affect the evolution of preferences as well as secondary sexual characters. A major advance in our understanding of sexual selection, and a major contribution of work on guppies, was the realization that mating preferences are not fixed but may show plasticity in their expression (section 5.4) or differ genetically within and among populations (sections 6.2, 6.6, and 6.7). Theoretical work on the Fisherian and good-genes processes of sexual selection has made it clear that mate choice and secondary sexual characteristics are likely to be evolutionarily linked through genetic correlations. Thus, changes in mating preferences do indeed affect male color patterns in guppies, as has long been assumed (Arrow 1: evolutionary effect), but changes in color patterns may also have evolutionary effects on mating preferences as a result of the correlation between the two traits (Arrow 4: evolutionary effect; section 6.7).

In addition to indirect selection on preferences described by good-genes and Fisherian models, recent discussions among theoreticians have focused on direct agents of selection on mate choice, and some empirical support for this idea is emerging from studies of guppies (sections 5.3, 5.4, and 5.5). Predators, parasites, and possibly the social environment may affect the direct costs and benefits of mate choice, leading to direct selection on preferences (Arrow 3: evolutionary effect). There is relatively little empirical data on the magnitude of costs of mate choice relative to direct benefits and effects of indirect selection. For example, predation is a clear risk to guppies in many populations, but the degree to which mate selection itself entails additional risks of predation or parasite infection or is affected in other ways by direct natural selection remains to be determined. The degree to which differences in costs or benefits of mate choice vary among populations and contribute to divergence in mating preferences among populations remains to be determined (but see sections 5.5 and 6.6 for some hints).

Although models for the evolution of mating preferences have been discussed at length from a theoretical point of view, empirical evidence has been difficult to obtain. We have some empirical information supporting both direct and indirect selection on preferences from guppies, but it is far from complete or conclusive. This information is valuable, however, because comparable data have been obtained from few other species.

In the same way that individual males can adjust the expression of secondary sexual traits, females show flexibility in the expression of their mating preferences (Arrow 3: plasticity). Females adjust their patterns of mate choice in response to the presence of predators and to aspects of their social experience (section 5.4), but effects of food availability and parasites on preferences have not been investigated.

The expression and possibly the evolution of mating preferences also appear to depend on how signals from males are received and processed by the sensory and cognitive systems (Arrow 5: plasticity and evolutionary effect; section 5.5). The visual signals received from males depend, in turn, on properties of ambient light, water as a transmission medium, and the reflectance of male color patterns (Arrow 6: plasticity). These physical properties of the visual environment may lead to selection on the sensory system (Arrow 6: evolutionary effect). In addition, selection on other functions of sensory systems can lead to evolutionary changes in mating preferences (Arrows 5 and 6: evolutionary effects). For example, requirements for finding food or detecting predators may shape the sensory system in ways that constrain or bias the evolution of mating preferences. We are only beginning to obtain empirical information about how mating preferences depend on sensory systems and the physics of signal transmission.

Despite the intensive study of female choice (intersexual selection) in guppies, male-male competition (intrasexual selection) has been somewhat neglected in work with guppies (section 3.4). Ironically, Haskins et al. (1961) actually concluded that color patterns were more likely to function in male-male competition than in female choice, but subsequent work has focused primarily on female choice. There is some suggestion that male color patterns may be related to success in intrasexual interactions (Arrow 7: evolutionary effect), but there is still a need for data on the relative importance of inter- and intrasexual selection in guppies.

The apparent simplicity of the guppy mating system has made this species attractive for studies of sexual behavior and mate choice and has made it ideal for testing some of the major ideas in sexual selection theory. The empirical testing of theoretical ideas continues, and numerous questions remain to be answered. Work on guppies has also uncovered a wealth of unsuspected behavioral complexity and flexibility. The guppy system will continue to provide a testing ground for major hypotheses derived from theory, but in addition it continues to provide an opportunity for empirical studies to uncover astonishing sophistication in the behavior of a small fish.

7.2 Questions for Future Research

My own work with guppies and my survey of the literature for this monograph have suggested a number of areas for further research. It is impossible, of course, to make an exhaustive list of future research areas in any field, and this is especially difficult for research on guppies because there are so many interesting questions, large and small, that remain to be answered. Here, I present several of the large questions that I think could provide profitable directions for future work with guppies. Some aspects of these questions are already in the beginning stages of investigation, others I have included specifically because they have received little attention. This list is undoubtedly idiosyncratic, and readers will probably add their own questions to the list and fail to see the significance of some of mine. See the text and summaries of chapters 3–6 for additional details and more suggestions for further research.

1. *What is the sensory basis for female mating preferences?* Work on the sensory and cognitive processing that gives rise to mating preferences in guppies has been productive, but is only beginning. We know that differences in mating preferences can result from very basic changes in the visual system of guppies. We do not yet know if variation in preferences between populations is the result of these kinds of changes. Although we understand how the physical features of the environment (e.g., water color) can affect the transmission of signals from males to females, we have very little direct information about how these signal transmission properties affect the evolution of communication systems. The possibility that selection on sensory systems in contexts other than mate choice (e.g., food finding) can affect the evolution of mating preferences also needs to be investigated. What are the biases of female guppies in response to colored objects other than males? Finally, work on sexual communication via sensory modalities other than vision (e.g., the lateral line; Rush, pers. comm.) has exciting possibilities.

2. *What are the behavioral rules females use in selecting their mates?* Despite theoretical interest, the rules used in mate choice are not well understood for guppies or any other species. The few studies of choice rules in guppies and other fish species suggest that it should not be hard to learn more. The problem needs to be approached with the mating system of the species in mind, rather than with the goal of testing purely theoretical predictions. In guppies, we need to use as a starting point the observation that males approach and court females rather than vice versa. Simple questions need to be addressed first—for example, whether or not the previous male to court a female affects her response to the current male. With an empirical foundation, issues raised by theoretical work can then be addressed.

3. *How do social interactions affect the sexual behavior of males and females?* Our appreciation for the complexity and flexibility of the reproductive behavior of male and female guppies comes especially from the realization that the social environment—the presence and behavior of other individuals—-plays an important role. This includes general influences of the social environment such as the effect of density on courtship behavior, and also very specific effects of behavioral interactions—for example, copying of mate choice by females. We know a little bit here and there about the effects of the social environment on sexual behavior of guppies, but more empirical work is needed before we can arrive at more general syntheses.

4. *What are the effects of male-male competition in a female-controlled system?* Guppies differ from related species of poeciliids and from species such as sticklebacks in that male-male competition seems to have relatively minor effects on the outcome of sexual selection. This is because the mating system is largely controlled by females. It would be interesting to know

more about why guppies differ from other species in this respect. We do know that male-male competition, aggressive interactions, and dominance relationships do play some role in the guppy mating system, but more work is needed to fully understand the interplay between inter- and intrasexual selection.

5. *Do male guppies show individual variation in mating strategies?* Guppies provide an example of the facultative use of alternative mating strategies. Most males switch between courtship displays and sneak copulation attempts, depending on the immediate situation. In other species, individuals are more likely to specialize in one or the other tactic, and in some species this behavioral polymorphism may have a genetic basis and can be related to body size. Again, we can ask why guppies differ from these other species. We also need to examine the possibility that male guppies differ, at least subtly, in their use of mating tactics, and to further examine the possible relationship of body size to courtship behavior and mating success.

6. *Do female mating preferences always lead to sexual selection on male traits?* As we saw in chapter 6, mating preferences may not always translate into sexual selection on male traits, and preferences are not similar in all populations. Further study of the factors that determine the relationships between mating preferences, mate choice, male mating success, and evolution of secondary sexual characters is needed in guppies and other species. We do not know whether mating preferences measured in laboratory studies always predict corresponding patterns of male mating success in wild population. For example, females may express their mating preferences facultatively, depending on the presence or absence of predators or other factors. The social environment may also affect the ability of females to mate with the males they prefer. For example, sneak copulations and interference among courting males could reduce the ability of females to choose, especially when population densities are high, sex ratios are male biased, or visibility of courting pairs to other males is good. Sneak copulations may also become more frequent in situations when males reduce their use of courtship displays. These ideas need to be tested in order to understand the relationship between mating preferences and the evolution of secondary sexual traits in more detail.

7. *Why does selection sometimes lead to genetic divergence and sometimes to adaptive plasticity?* The ability of males and females to adjust their sexual behavior depending on whether or not predators are present appears to be adaptive in that it minimizes the risk of predation. There is also genetic divergence among populations corresponding to differences in risk of predation. The sexual behavior of guppies therefore appears to have responded to selection, by predators in this example, both genetically and through plasticity of behavior. This should be a good system in which to

learn more about the conditions under which adaptive plasticity evolves. Why, for example, does adaptive plasticity evolve for some traits, while others show genetic variation among populations? What determines the degree of flexibility shown in a given behavioral trait?

8. *How do the theoretical models of sexual selection apply to guppies?* A number of studies on guppies were initiated with the goal of testing the once highly controversial good-genes and Fisherian models of sexual selection. These studies have made significant empirical contributions in this area with evidence in support of both mechanisms. The controversy has now largely been laid to rest with the realization that "good genes" and Fisherian "runaway" refer to mechanisms of indirect selection on mating preferences that are not mutually exclusive and may often, if not always, operate together. Nevertheless, the empirical work spawned by this theoretical controversy has shed a great deal of light on the evolutionary dynamics of sexual selection in guppies. The guppy system still has a great deal of potential to further illuminate these theoretical issues. For example, although carotenoid colors are clearly condition dependent in guppies, no study has demonstrated a relationship of carotenoids to fitness traits in offspring, as predicted by the good genes model. It should be possible to test for heritable effects associated with ability to cope with parasite infection, food limitation, and other stresses that affect the expression of carotenoid colors. Another question worth addressing with guppies concerns the degree to which genetic variances and covariances of mating preferences and secondary sexual characters differ among populations. The genetic structure of these traits is central to all models of sexual selection, and comparative studies could be especially informative.

9. *Why are guppy color patterns so polymorphic?* Surprisingly, the most striking aspect of guppy populations, the extreme genetic polymorphism in the color patterns of males, remains essentially unexplained. Further studies of the inheritance and expression of color patterns could help us understand the genetic origins of the polymorphism. What, for example, are the roles of mutation and recombination in generating new color patterns? Does selection affect the degree of polymorphism? Behavioral studies could also be informative. Do patterns of mate choice contribute to the maintenance of this variation? There is some suggestion that frequency-dependent mate choice might contribute to the extreme degree of polymorphism in color patterns, but this remains unconfirmed. A combination of further experimentation on frequency-dependent mate choice, a better understanding of the genetics of color patterns, and theoretical models might provide answers to the polymorphism puzzle.

10. *Do divergent guppy populations provide a model for speciation?* We have evidence that mating preferences of females have diverged among guppy populations, some evidence that females prefer males from their

own population, and some evidence for assortative mating between populations. In order to understand the role sexual selection plays in speciation, more studies need to investigate the possibility of incipient isolation among guppy populations. Do female preferences for native males result in assortative mating? Do males discriminate between females from their own and different populations? What traits are important in population discrimination?

7.3 Concluding Remarks

The guppy is a small fish, studied by a small (but growing) community of researchers. Yet the insights provided by research on guppies have been anything but small. Studies of guppies have provided much-needed empirical support and critical tests of theories and hypotheses and have suggested new directions for theory and empirical studies of other species. Not every question posed has been answered, and not every hypothesis has proved to be testable, even with guppies. Furthermore, in every study, many more questions have been raised than have been answered. I hope that some of the questions I have listed above will provide a point of departure for yet more illuminating insights from guppies. Despite the fact that they are now the subject of a monograph, I expect that guppies will continue to be a source of fascination and productive study in the years to come.

Appendix
Experimental Methods: How to Build a Better Bordello

This appendix describes some of the methods used in experiments on sexual selection in guppies and discusses their usefulness and weaknesses. Many of the basic designs are common to studies with other systems; here I will emphasize how specific features of guppy biology can be accommodated or exploited by researchers. Guppies are especially useful for sexual selection studies because female mating preferences can be inferred from the behavior of females themselves rather than from male mating success, which may be affected by male-male competition.

Much of the strength of the data that have been gathered on female choice in guppies stems from the fact that similar results have been obtained by different researchers using different methods. The various experimental designs each have strengths and weaknesses, but no one approach can be said to be better than another. At worst, discrepancies among different experiments may be attributable to differences in design.

A.1 Design of Mate Choice Experiments

The designs of sexual selection experiments for guppies differ in the physical plan of the aquaria used, in the social groupings of experimental fish, in the kinds of data obtained, and in the specific questions that can be addressed. I have categorized the experimental designs into open aquarium experiments, dichotomous choice experiments, and single male tests (fig. A.1). These differ in the data used to measure female preferences, the degree to which females can compare males, the degree to which male-male competition can occur, and in other aspects.

Open Aquarium

The open-aquarium design (e.g., Houde 1987, 1988a,b; Houde and Endler 1990; Endler and Houde 1995) involves conducting observations on groups of fish that are free to interact throughout an aquarium. This design is probably the most realistic way of replicating natural social groupings of

A. Open Group

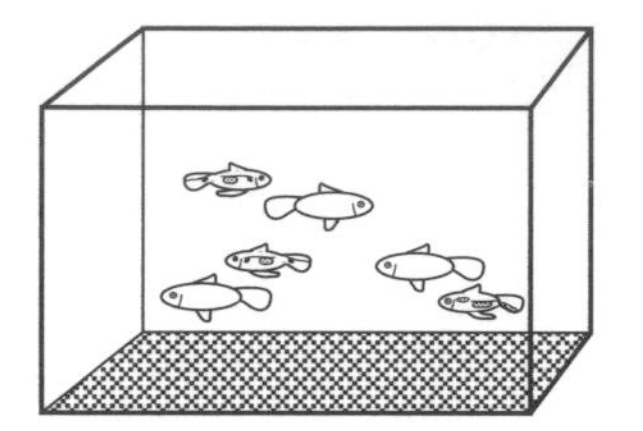

B. Divided Aquarium

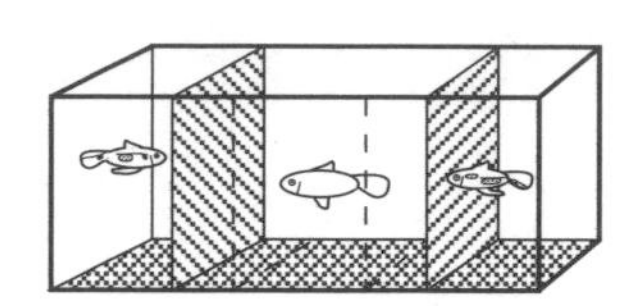

C. One on One

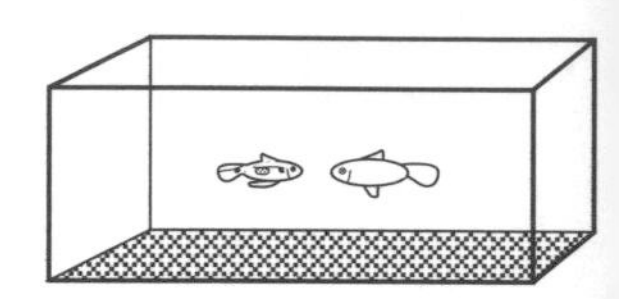

Figure A.1 Three experimental designs for measuring female choice in guppies.

guppies in the laboratory. It can be used for mate choice experiments and for investigations of patterns of male courting behavior simultaneously. A potential disadvantage is that male-male competition can potentially affect results on female choice, but it can also be observed and measured. In general, the aquarium is bare except for a layer of gravel. Gravel is necessary for normal swimming and foraging activity; without it the fish tend to press against the bottom of the tank and swim in an agitated manner. In some cases it makes sense to place partial partitions in the aquarium so that fish can get away from one another, especially when experimental groups are large. Otherwise, males sometimes have difficulty courting females without interruptions from other males. In the wild, males often have the opportunity to court females behind a rock or in another place that is visually isolated from other guppies.

For female choice experiments, the relative attractiveness of the males in the experimental group can be measured by observing the sexual responses of females to each male. An observation session usually consists of a series of focal samples of the courtship of the males. Males can be individually identified by their color patterns (chosen to be easily distinguishable) using a previously prepared identification chart showing the "field marks" of each male. In the focal observations, each male is observed for a set time interval (usually 5 or 10 min), in random order. To obtain female choice data, each courtship display of the focal male is noted along with the female's response.

For every male display, we can score the degree of response of the female in terms of how much of the courtship sequence described above is completed. For example, in experimental observations, I have used a scoring system in which a male's display receives a score of 0 if the female does not respond, and a score of 5 if the female sustains her response through the entire courtship sequence to copulation (table A.1). In my own experiments, I record a sexual response to a male's display if the female actually glides toward him (score 2 or above; table A.1), and no response if she

Table A.1

Scoring System for Female Responses to Male Displays

Score	Female and Male Behavior
0	No response; female ignores male
1	Female orients toward male but does not move closer
2	Female glides toward male
3	Male and female circle around each other
4	Copulation attempt; mate thrusts and makes gonopodial contact
5	Copulation; gonopodial contact followed by male jerking

SOURCE: Endler and Houde 1995.

NOTES: Each stage includes behavior from all previous stages, plus behavior noted. Males sometimes omit Stage 3 and attempt copulation as soon as the female responds.

ignores him (score 0) or merely starts to orient toward him (score 1). A male's ability to elicit responses from females (e.g., the fraction of his displays that elicit a response or the mean response score for his displays) can be used as a measure of his attractiveness. The use of these measures in female choice experiments is described in chapter 3.

The mean or median response score, or the fraction of displays that elicit a response by the above criteria, then gives a measure of attractiveness for each male. This measure represents a composite of the response behavior of all the females in the experimental group unless an attempt is made to identify the females individually (difficult in practice). Male attractiveness can then be related to color pattern and other characteristics statistically. A potential problem is that if some females receive more courtship than others, the composite attractiveness measure could be weighted toward the preferences of these females. This potential source of bias can be minimized within observation sessions by trying to match the females as closely as possible in size, age, and prior experience.

Because pregnant female guppies (i.e., all females that have been housed with males) tend to be completely unresponsive to males, most researchers use virgin females in choice experiments, although non-virgins can provide useful data (e.g., Dugatkin and Godin 1992b, 1993). But virgins that have matured without ever seeing a male are likely to mate very soon after being introduced to males. For open-aquarium experiments, introducing completely naive virgins results in many copulations in a short period of time, and in refractory males that do not court females. One solution to this problem has been to introduce the virgin females to the males in the experimental aquarium the morning before the morning of the observation session. After twenty-four hours, the females are still responsive to males, but they are less likely to copulate during an observation session, although some matings still occur. Observations based on these "recently virgin" females provide good data on preferences that reflect the actual mating propensities of the females (Houde 1988b).

It is also possible to measure sexual selection more directly in an open-aquarium experiment by scoring actual copulations with virgin females or by estimating male mating success. The behavioral measures of preference can be validated if female responses and actual matings are scored in the same experiment (e.g., Bischoff et al. 1985; Houde 1988b; Kodric-Brown 1992). Using actual matings to measure preference requires large numbers of virgin females, because one female is needed per mating, and it is desirable to observe numerous matings in each group of competing males (three to five times the number of matings as there are males in the group would be ideal). Another approach, which can avoid the potential problem that naive virgin females may be relatively indiscriminate in their first matings, is to use paternity tests to determine the mating success of males in long-term social groupings in open aquaria. Male color patterns have very high heritability and can be used to identify the fathers of offspring born to females in experimental groups, providing an estimate of mating success (e.g., Farr 1980b; Houde 1988b; see chapter 5). Data on mating success can then be related statistically to behavioral data. The process of isolating females and rearing their offspring can be lengthy and requires many aquaria.

In order to observe normal courtship sequences, the density of fish and the sex ratio need to be within the natural range of variation. When densities are too high or the sex ratio is male biased, males tend to interfere with one another's courtship more than normal. Under these conditions, male-male competition may become more important and females may have fewer opportunities to exercise their preferences. It is generally best to use an even or female-biased sex ratio (up to two females per male), and a density of no more than 10–15 males per square meter of aquarium bottom.

Another difficulty that becomes apparent in this kind of experiment is that actual copulations that occur during observation sessions can interfere with data collection. This is because males, and sometimes females, can have long refractory periods after copulating during which they do not engage in sexual behavior. Males become lethargic and even sick-looking after mating and sometimes do not court females again for as much as one hour. This can be a big problem for conducting observations!

Male-male competition can be scored in an open-aquarium situation at the same time as female choice is being scored. In addition to recording the responses of females to the focal male, interactions of other males with the focal male can be recorded. Overt aggression and agonistic displays are uncommon and may be a sign that females are behaving abnormally. Aggression also seems to be fairly common when there are only two males in the group, as in the case in a number of studies where aggression was observed (e.g., Farr 1980b; Kodric-Brown 1992). Aside from overt agonistic

Table A.2

Example of a Method for Scoring Male-Male Competition in Focal Male Observations

Length of Interaction	Focal Male	Other Male	Outcome: Focal Male . . .	Code
Short	With female	Intruder	**F**ends off intruder	**F**
	With female	Intruder	Has female take**N**-over by intruder	**N**
	Intruder	With female	Is fende**D** off by other male	**D**
	Intruder	With female	**T**akes-over female	**T**
Long	With female	Intruder	**K**eeps female	**K**
	With female	Intruder	**L**oses female	**L**
	Intruder	With female	**W**ins female	**W**
	Intruder	With female	Is **U**nsuccessful	**U**

SOURCE: Houde 1988b.

NOTES: The focal male is the male that is with the female when the interaction begins. Short interactions are very brief; long interactions involve males jockeying for position for several seconds, possibly while they are in pursuit of the female who has fled. Interactions are categorized according to whether the focal male was originally with the female, and whether he is with the female after

behavior, interactions between males occur when one male attempts to court a female already being courted by another male (see chapter 2). Table A.2 gives an example of how these interactions can be scored.

Dichotomous Choice

The dichotomous-choice design is the fish version of a dichotomous-choice design used in many other studies of female choice. Other examples include phonotaxis experiments on crickets and anurans (e.g., Hedrick 1986; Gerhardt 1987; Bailey et al. 1990; Ryan et al. 1990b), choice experiments on red jungle fowl (Zuk et al. 1990a,b,c), and choice experiments with other fish (e.g., swordtails: Basolo 1990a,b, 1995a,b). This design uses side-association data based on the time spent by the female in proximity to different males or different stimuli. It does not depend on identification of specific sexual behaviors of females.

The dichotomous-choice design is more manipulative and thus more artificial than open aquarium observations. It does allow a greater degree of control over the experiment, particularly of male-male competition, though. The basic method is to divide an aquarium into three chambers and to place a male in the two end chambers and a female in the center (fig. A.1). The female is thus in a position to view and compare the two males, and her preference can be assayed through behavioral observations. Virgin females or females that have recently given birth to young are usually used since they are most likely to show a sexual interest in the males. If the aquarium is partitioned with plain glass or clear plastic panels, the males have at least the potential to see one another and influence one another's behavior. This can be controlled by using a polarizing filter on each partition, with the plane of polarization of one partition rotated 90 degrees relative to that of the other partition. This way, the female can see both males, each male can see the female, but the males cannot see each other. Because polarizing films usually reduce light transmission, it may be necessary to illuminate each male's compartment separately from the female's compartment using low wattage bulbs. One study (Bischoff et al. 1985) used one-way glass partitions so that the males could not see the females. This avoids feedback between female behavior and male behavior, and the authors report that the males displayed spontaneously. In most guppy strains and populations, males need to see the female to display, so it is probably better to use the polarizing technique. A general disadvantage of experimental aquaria with dividers is that pheromonal communication is reduced or prevented, which may reduce the responsiveness of males and females to each other.

Female preference in dichotomous-choice experiments is usually measured through side-association data—some measure of the relative amount of time spent near each male. In order to be confident that this associative behavior is related to mate choice, it is important to use receptive females in the experiments. In most studies, the female's center chamber of the aquarium is divided into three or sometimes five equal sections, with lines drawn on the front glass. The female is initially placed in the center section and allowed to acclimate. Side-association data are collected during a set observation period (e.g., 15 min) either by recording the total amount of time the female spends in each section of her chamber or by recording which section she is occupying in point samples (e.g., every 30 sec). To eliminate the possibility that the female is attracted to a particular side of the aquarium and not the male on that side, most experiments include a second observation session for each pair of males in which the positions of the two males are reversed.

A measure of preference is then calculated using the amount of time the female spends in the sections nearest the males. Even though this eliminates time not spent near either male (i.e., in the center section), the time spent near one male is still not statistically independent of the time spent near the other male. Preference must therefore be measured as the difference in time spent with each male (i.e., one, not two, data points per pair of males). This difference is usually standardized by the total amount of time spent in the sections nearest the males. If A is the time spent in the section nearest male A, and B is the time spent nearest male B, then a measure of relative preference is (A–B)/(A+B) (e.g., Stoner and Breden 1988).

The main criticism of dichotomous-choice studies is that side-association data do not represent direct observations of the sexual behavior of females. There are now enough studies using side-association data that have produced results consistent with other kinds of studies that this should no longer be a serious concern. In particular, a few studies have obtained consistent results from both side-association data and other more direct measures of female choice and male mating success in the same experiment (Bischoff et al. 1985; Houde and Torio 1992; Kodric-Brown 1992).

Even using a dichotomous-choice setup, more direct data on mate choice can be obtained. Males display to females through the partitions, and females exhibit the characteristic sexual responses to their displays. Thus, the fraction of displays eliciting a response can be used as a measure of preference as in open-aquarium experiments (Houde and Torio 1992). There is probably a greater degree of independence in the data for each male than with side-association data, but it is still probably safest to calculate a single measure of preference for each pair of males rather than for each male separately.

A difficulty with dichotomous-choice experiments, especially with side-association data, is that the female may not sample both males during the observation session. A female that moves near one male initially and never visits the other male may or may not have made a choice. If she is choosing, she should move to the same male when the males' positions are switched. One way to deal with this problem is simply to discard any observations in which the female does not visit both males, or to discard observations in which she is not courted by both males.

For specific experiments, a number of variations on the dichotomous-choice design are possible. Dugatkin (1992a) has explored the possibility that females copy the mate choice of other females by allowing his females to see another female placed near one of the two males in a dichotomous choice (see chapter 6 for more details and references). This is a clever way to test this idea, although there is some question as to whether female guppies are likely to observe the association preferences of other females in natural situations.

Single-Male Tests

A final type of design for sexual selection experiments is one-on-one encounters between a single male and one or more females (e.g., Houde 1987; Reynolds and Gross 1992; Brooks and Caithness 1995d; Brooks 1996a,b). This design eliminates male-male competition, but allows free interaction between the male and the female. The experimental design involves placing one male in an aquarium with a single, usually virgin, female. This is similar to the open-aquarium design, but male-male competition cannot occur and interpretation of the data is somewhat different.

Data that can be obtained include aspects of the courtship behavior of males (e.g., frequency of displays and duration of courtship bouts), responses of the female to the male's displays, time to copulation, or whether or not copulation occurs within a specified time interval. In one of my experiments (Houde 1987), about half of the virgin females presented to males copulated within fifteen minutes. In practice, however, virgin females' inclination to copulate during observation sessions seems to depend on many factors, such as lighting, age of the female, and time of day or time since feeding.

Since only one male is present, the data from single male tests describe aspects of mating propensity rather than comparative choice. In other words, the intrinsic responses of a female to a particular male are measured, and the possibility of comparisons between males is eliminated. This could actually be useful in trying to determine the importance of relative or conditional rules for mate choice compared with intrinsic preferences (see chapter 5).

Strengths and Weaknesses

Each of the experimental designs for studying female choice in guppies has different strengths and weaknesses that depend on the specific goals of the experiment. The actual design chosen may depend on uncontrollable factors such as the availability of different males, the availability of virgin females, and the time required for observation trials.

Single-male tests and dichotomous-choice experiments have the advantage that females are being tested individually. They differ in that single-male tests assess the intrinsic pattern of female responses to each male without an opportunity for comparison, while other designs allow females to compare males and assess their relative preference for different males. Thus dichotomous-choice tests are not a good design for reconstructing the precise preference functions of females because a female's response to one male may depend as much on the phenotype of the other male as on the male in question.

The greatest statistical power for detecting a preference for a particular male character is obtained by testing as many males as possible, each with only one virgin female. In this case, individual variation among females adds to the statistical error, however. This error can be reduced by testing a given male or pair of males with many different females, but at a cost in terms of statistical power.

Different experimental designs for female choice experiments are more or less sensitive to individual variation in preferences in different ways. In open-aquarium experiments with fairly large experimental groups, the average responsiveness of the females in the group for each of the males is measured, which tends to eliminate some of the effects of individual variation (e.g., Houde 1987). By marking the females in the group (e.g., with colored ink tattoos), individual variation in preferences can be assessed (Houde, unpublished data), but the observations necessary to compare individual females are cumbersome.

Thus, in choosing a design for mate choice studies, the goals of the study, the availability of males and females, and the time needed for the actual tests all need to be weighed carefully.

A.2 Measuring Male Mating Success

Given the reproductive biology of guppies (see chapter 2), there are a number of methods for measuring the actual mating success of males. First, copulations can be observed directly by allowing two or more males to compete for matings with virgin females. Unlike some bird species (e.g., jungle fowl: Zuk et al 1990a,b), male guppies cannot be tethered in mate choice experiments, so male-male interactions cannot be eliminated when observations of actual matings are needed. The interpretation of results is relatively straightforward if two males compete and only one mates with the female, although male-male competition may be artificially intensified because of the biased sex ratio (see discussion in chapter 3). When several males are competing for matings with several females simultaneously, the male refractory period between matings (see discussion of open-aquarium design in sec. A.1) may give all males more equal opportunities to mate, thus reducing the overall variance in mating success. Direct observation of copulations may also give unreliable results if naive females are mating indiscriminately (see discussion of open-aquarium design).

It is also possible to assess mating success more accurately by determining the proportion of sperm in a recently inseminated female attributable to competing males. This can be done by radioactively labeling the sperm of males as described by Luyten and Liley (1991). This technique could be especially useful in determining mating success or paternity when sneak copulations are frequent.

Another, somewhat less direct way to measure male mating success is to allow males and females to interact freely and then to isolate the females and reconstruct the distribution of matings by determining paternity of offspring. If fish are maintained in the aquaria for at least six weeks before isolating the females, the problem of indiscriminate mating by virgins can be reduced because the females will all have gone through two or more receptive periods in contact with males (see Houde 1988b for details). Paternity can be scored using the color patterns of sons as genetic markers (see chapters 1 and 3; Haskins et al. 1961; Farr 1980b; Houde 1988b; Kodric-Brown 1993; Brooks 1996b). Molecular methods (e.g., DNA fingerprinting) should also be feasible for determining paternity (P. Parker, pers. comm.; Houde, unpublished data; Foo et al. 1995; Hornaday et al. 1995; Sato et al. 1996) and could be more efficient than using color pattern markers.

A.3 Measuring Male Color Patterns

The goal of many sexual selection studies is to investigate female mating preferences for particular male characteristics, and to infer patterns of sexual selection. The basic approach is to analyze the effect of male traits on mating success or surrogate measures of attractiveness obtained in behavioral observations. Depending on the nature of the data, regression or multiple-regression techniques are most commonly used. For quantitative methods and examples, see Arnold and Wade (1984a,b), Endler (1986), Schluter (1988), and Endler and Houde (1995).

In trying to infer what male traits could be used as cues in female choice, there are an infinite number of possibilities. For guppies, measures of the color pattern, body size and shape, and courtship behavior have been used. Measuring body size and shape is relatively straightforward. Measurements of anaesthetized fish can be made directly or from photos or video images. Courtship behavior is usually quantified in terms of the frequency of display per unit time with direct observation. A few studies have measured the duration of individual sigmoid displays from videos or films. Problems with inferring female preferences for male behavioral elements are discussed in chapter 3.

The majority of studies of female choice in guppies have focused on the role of male color patterns as cues. The simplest color pattern characters include the numbers and sizes of spots, and total area of particular colors. These are easily measured from photographs or video images using some sort of computer digitizer. Areas of colors can be presented as an absolute measure (e.g., square millimeters), or can be standardized by dividing by body area (fraction of body area). The measurement of actual spot colors is more difficult. The theory and practicalities of measuring color in guppies and other natural systems have been developed by Endler (1990, 1991, 1993; see also Bennett et al. 1994; Zuk and Decruyenaere 1994). Each color pattern spot reflects a particular spectrum of light, which depends on the spectrum of incident light and the transmission spectrum of the water. Data on individual spots can then be combined into composite measures of total colored area, average brightness or variation in brightness of the color pattern, average chroma and variation in chroma, and so on. Relating more complex measures of the appearance of color patterns to female preferences could give some insight into the cognitive processing behind mate choice.

In practice, the measurement of color pattern characters depends on the technology available. At the simplest level, color classes (e.g., black, orange, green, yellow) are simply categorized by eye. Variation within color classes can be done crudely by comparison with Munsell chips (Munsell 1976; Houde and Torio 1992). More quantitatively, it is possible to measure the relative brightness of particular colors in guppy spots using a densitometer and a system of filters that limit the frequency of incident light (Kodric-Brown 1989, Nicoletto 1991). Reflectance spectra can be estimated more directly by illuminating spots with monochromatic light at a full series of wavelengths and measuring the reflected light at each wavelength using a radiance sensor (see Endler 1990, 1991 for full details). The technology required for the latter method is expensive, however.

References

Abrahams, M. V. 1989. Foraging guppies and the ideal free distribution: The influence of information on patch choice. *Ethology* 82: 116–126.

Abrahams, M. V. 1993. The trade-off between foraging and courting in male guppies. *Anim. Behav.* 45: 673–681.

Andersson, M. 1994. *Sexual Selection*. Princeton University Press, Princeton, N.J.

Andersson, S. 1989. Sexual selection and cues for female choice in leks of Jackson's widowbird *Euplectes jacksoni*. *Behav. Ecol. Sociobiol.* 25: 403–410.

Andersson, S. 1991. Bowers on the savanna: Display courts and mate choice in a lekking widowbird. *Behav. Ecol.* 2: 210–218.

Andersson, S. 1992. Female preference for long tails in lekking Jackson's widowbirds: Experimental evidence. *Anim. Behav.* 43: 379–388.

Angus, R. A. 1989. A genetic overview of poeciliid fishes. In G. K. Meffe and F. F. Snelson, eds., *Ecology and Evolution of Livebearing Fishes (Poeciliidae)*, pp. 51–68. Prentice Hall, Englewood Cliffs, N.J.

Archer, S. N., and J. N. Lythgoe. 1990. The visual pigment basis for cone polymorphism in the guppy, *Poecilia reticulata*. *Vision Research* 30: 225–233.

Archer, S. N., J. A. Endler, J. N. Lythgoe, and J. C. Partridge. 1987. Visual pigment polymorphism in the guppy *Poecilia reticulata*. *Vision Research* 27: 1243–1252.

Arnold, S. J., and M. J. Wade. 1984. On the measurement of natural and sexual selection: Theory. *Evolution* 38: 709–719.

Austad, S. N., and R. D. Howard. 1984. Introduction to the symposium: Alternative reproductive tactics. *Am. Zool.* 24: 309–319.

Baer, C. F., M. Dantzker, and M. J. Ryan. 1995. A test for preference of association on a color polymorphic Poeciliid Fish: Laboratory study. *Environ. Biol. Fish* 43: 207–212.

Baerends, G. P., R. Brouwer, and H. T. Waterbolk. 1955. Ethological studies on *Lebistes reticulatus* (Peters), I. An analysis of the male courtship pattern. *Behaviour* 8: 249–334.

Bailey, W. J., R. J. Cunningham, and L. Lebel. 1990. Song poser, spectral distribution and female phonotaxis in the bushcricket *Requena verticalis* (Tettigoniidae: Orthoptera): Active female choice or passive attraction? *Anim. Behav.* 40: 33–42.

Bakker, T.C.M. 1990. Genetic variation in female mating preferences. *Netherl. J. Zool.* 40: 617–642.

Bakker, T.C.M. 1993. Positive genetic correlation between female preference and preferred male ornament in sticklebacks. *Nature* 363: 255–257.

Bakker, T.C.M. 1994. Evolution of aggressive behavior in the threespine stickleback. In M. A. Bell and S. A. Foster, eds., *The Evolutionary Biology of the Threespine Stickleback*, pp. 345–380. Oxford University Press, Oxford.

Bakker, T.C.M., and M. Milinski. 1991. Sequential female choice and the previous male effect in sticklebacks. *Behav. Ecol. Sociobiol.* 29: 205–210.

Bakker, T.C.M., and B. Mundwiler. 1994. Female mate choice and male red coloration in a natural three-spined stickleback (*Gasterosteus aculeatus*) population. *Behav. Ecol.* 5: 74–80.

Bakker, T.C.M., and A. N. Pomiankowski. 1995. The genetic basis of female mate preferences. *J. Evol. Biol.* 8: 129–171.

Bakker, T.C.M., and P. Sevenster. 1983. Determinants of dominance in male sticklebacks (*Gasterosteus aculeatus L.*). *Behaviour* 86: 55–71.

Ballin, P. J. 1973. Geographic variation of courtship behavior in natural populations of the guppy Poecilia reticulata (Peters). M.Sc. thesis, University of British Columbia, Vancouver.

Balmford, A., I. L. Jones, and A.L.R. Thomas. 1994. How to compensate for costly sexually selected tails: The origin of sexually dimorphic wings in long-tailed birds. *Evolution* 48: 1062–1070.

Balmford, A., and A. F. Read. 1991. Testing alternative models of sexual selection through female choice. *TREE* 6: 274–276.

Barlow, J. 1992. Nonlinear and logistic growth in experimental populations of guppies. *Ecology* 73: 941–950.

Basolo, A. L. 1990a. Female preference predates the evolution of the sword in swordtail fish. *Science* 250: 808–810.

Basolo, A. L. 1990b. Female preference for male sword length in the green swordtail, *Xiphophorus helleri* (Pisces: Poeciliidae). *Anim. Behav.* 40: 332–338.

Basolo, A. L. 1995a. A further examination of a preexisting bias favoring a sword in the genus *Xiphophorus. Anim. Behav.* 50: 365–375.

Basolo, A. L. 1995b. Phylogenetic evidence for the role of a preexisting bias in sexual selection. *Proc. Roy. Soc. Lond. B* 259: 307–311.

Bateman, A. J. 1948. Intra-sexual selection in *Drosophila. Heredity* 2: 349–368.

Bateson, P. 1978. Sexual imprinting and optimal outbreeding. *Nature* 273: 659–660.

Bateson, P. 1983. *Optimal Outbreeding*. Cambridge University Press, Cambridge, U.K.

Beaugrand, J. P., J. Caron, and L. Comeau. 1984. Social organization in small heterosexual groups of green swordtails (*Xiphophorus helleri*, Pisces, Poeciliidae) under conditions of captivity. *Behaviour* 91: 24–60.

Bennett, A.T.D., I. C. Cuthill and K. J. Norris. 1994. Sexual selection and the mismeasure of color. *Am. Nat.* 144: 848–860.

Berglund, A. 1993. Risky sex: Male pipefishes mate at random in the presence of a predator. *Anim. Behav.* 46: 169–175.

Bildsøe, M. 1988. Aggressive, sexual, and foraging behaviour in *Poecilia velifera* (Pisces: Poeciliidae) during captivity. *Ethology* 79: 1–12.

Bisazza, A. 1993. Male competition, female mate choice and sexual size dimorphism in Poeciliid fishes. *Mar. Behav. Physiol.* 13: 257–286.

Bisazza, A., and G. Marin. 1991. Male size and female mate choice in the eastern mosquitofish (*Gambusia holbrooki*: Poeciliidae). *Copeia* 1991: 730–735.

Bischoff, R. J., J. L. Gould, and D. I. Rubenstein. 1985. Tail size and female choice in the guppy (*Poecilia reticulata*). *Behav. Ecol. Sociobiol.* 17: 253–255.

Boake, C.R.B. 1985. Genetic consequences of mate choice: A quantitative genetic method for testing sexual selection theory. *Science* 227: 1061–1063.

Boake, C.R.B. 1986. A method for testing adaptive hypotheses of mate choice. *Am. Nat.* 127: 654–666.

Boake, C.R.B. 1989. Repeatability: Its role in evolutionary studies of mating behavior. *Evol. Ecol.* 3: 173–182.

Borgia, G. 1979. Sexual selection and the evolution of mating systems. In M. S. Blum and N. A. Blum, eds. *Sexual Selection and Reproductive Competition in Insects*, pp. 19–80. Academic Press, New York.

Borgia, G. 1980. Sexual competition in *Scatophaga stercoraria*: Size- and density-related changes in male ability to capture females. *Behaviour* 75: 185–205.

Borgia, G. 1981. Mate selection in the fly *Scatophaga stercoraria*: Female choice in a male-controlled system. *Anim. Behav.* 29: 71–80.

Borgia, G. 1982. Experimental changes in resource structure and male density: Size-related differences in mating success among male *Scatophaga stercoraria*. *Evolution* 36: 307–315.

Borgia, G. 1985. Bower quality, number of decorations and mating success of male satin bowerbirds (*Ptilonorhynchus violaceus*): An experimental analysis. *Anim. Behav.* 33: 266–271.

Borgia, G. 1995. Why do bowerbirds build bowers? *American Scientist* 83: 542–547.

Borgia, G., and K. Collis. 1989. Female choice for parasite-free male satin bowerbirds and the evolution of bright male plumage. *Behav. Ecol. and Sociobiol.* 25: 445–454.

Borgia, G., and K. Collis. 1990. Parasites and bright male plumage in the satin bowerbird (*Ptyilonorhynchyus violaceus*). *Am. Zool.* 30: 279–285.

Borowsky, R. L. 1973. Social control of adult size in males of *Xiphophorus variatus*. *Nature* 245: 332–335.

Borowsky, R. L., and K. D. Kallman. 1976. Patterns of mating in natural populations of *Xiphophorus* (Pisces: Poeciliidae). I. *X. maculatus* from Belize and Mexico. *Evolution* 30: 693–706.

Bowden, B. S. 1969. A new method for obtaining precisely timed inseminations in viviparous fishes. *Prog. Fish Cult.* 31: 229–230.

Bradbury, J. W., and M. B. Andersson. 1987. Sexual selection: Testing the alternatives. Wiley, New York.

Breden, F. 1988. Sexual selection and predation risk in guppies. Reply. *Nature* 332: 594.

Breden, F., and K. Hornaday. 1994. Test of indirect models of selection in the Trinidad guppy. *Heredity* 73: 291–297.

Breden, F., and G. Stoner. 1987. Male predation risk determines female preference in the Trinidad guppy. *Nature* 329: 831–833.

Breden, F., M. Scott, and E. Michell. 1987. Genetic differentiation for anti-predator behavior in the Trinidad guppy, *Poecilia reticulata*. *Anim. Behav.* 35: 618–620.

Breden, F., H. C. Gerhardt, and R. K. Butlin. 1994. Female choice and genetic correlations. *TREE* 9: 343.

Breden, F., D. Novinger, and A. Schubert. 1995. The effect of experience on mate choice in the Trinidad guppy, *Poecilia reticulata*. *Environ. Biol. Fish* 7: 323–328.

Briggs, S. E., J.-G. J. Godin, and L. A. Dugatkin. 1996. Mate-choice copying under predation risk in the Trinidadian guppy (*Poecilia reticulata*). *Behav. Ecol.* 42: 151–157.

Brooks, R. 1996a. Melanin as a visual signal amplifier in male guppies. *Naturwissenschaften* 83: 39–41.

Brooks, R. 1996b. Sexual selection by female choice in guppies ((*Poecilia reticulata*). Ph.D. diss., University of Witwatersrand, South Africa.

Brooks, R. In press. Copying and the repeatability of mate choice. *Behav. Ecol. and Sociobiol.*

Brooks, R., and N. Caithness. 1995a. Does a male's attractiveness to a female depend on her previous experience? *S.A.J. Science* 91: 156–158.

Brooks, R., and N. Caithness. 1995b. Female choice in a feral guppy population: Are there multiple cues? *Anim. Behav.* 50: 301–307.

Brooks, R., and N. Caithness. 1995c. Female guppies use orange as a mate choice cue: A manipulative test. *S.A.J. Zool.* 30: 200–201.

Brooks, R., and N. Caithness. 1995d. Manipulating a seemingly non-preferred male ornament reveals a role in female choice. *Proc. Roy. Soc. Lond. B* 261: 7–10.

Brosset, A., and D. Lachaise. 1995. Evolution as a lottery conflicting with evolution via sexual selection in African rain forest-dwelling killifishes (Cyprinodontidae, Rivulinae, *Diapteron*). *Evol. Biol.* 28: 217–264.

Bruce, K. E., and W. G. White. 1995. Agonistic relationships and sexual behaviour patterns in male guppies, *Poecilia reticulata*. *Anim. Behav.* 50: 1009–1021.

Buchholz, R. 1995. Female choice, parasite load and male ornamentation in wild turkeys. *Anim. Behav.* 50: 929–943.

Burley, N. 1981. Mate choice by multiple criteria in a monogamous species. *Am. Nat.* 117: 515–528.

Bushmann, P. J., and J. R. Burns. 1994. Social control of male sexual maturation in the swordtail characin, *Chorynopoma riisei*. *J. Fish Biol.* 44: 263–272.

Cade, W. 1975. Acoustically orienting parasitoids: Fly phonotaxis to cricket song. *Science* 190: 1312–1313.

Carvalho, G. R., P. W. Shaw, A. E. Magurran, and B. H. Seghers. 1991. Marked genetic divergence revealed by allozymes among populations of the guppy *Poecilia reticulata* (Poeciliidae), in Trinidad. *Biol. J. Linn. Soc.* 42: 389–405.

Carvalho, G. R., P. W. Shaw, L. Hauser, B. H. Seghers, and A. E. Magurran. 1996. Artificial introductions, evolutionary change and population differentiation in Trinidadian guppies (*Poecilia reticulata*, Poeciliidae). *Biol. J. Linn. Soc.* 57: 219–234.

Chivers, D. P., B. D. Wisenden, and J. F. Smith. 1995. Predation risk influences reproductive behaviour of Iowa darters, *Etheostoma exile* (Osteichthyes, Percidae). *Ethology* 99: 278–285.

Clark, A. G. 1987. Natural selection and Y-linked polymorphism. *Genetics* 115: 569–577.

Clark, E., and L. R. Aronson. 1951. Sexual behavior in the guppy, *Lebistes reticulatus* (Peters). *Zoologica* 36: 49–66.

Clayton, D. H. 1990. Mate choice in experimentally parasitized rock doves: Lousy males lose. *Am. Zool.* 30: 251–262.

Clayton, D. H. 1991. The influence of parasites on host sexual selection. *Paras. Today* 7: 329–334.

Clutton-Brock, T. H., and G. A. Parker. 1992. Potential reproductive rates and the operation of sexual selection. *Q. Rev. Biol.* 67: 437–456.

Clutton-Brock, T. H., and A.C.J. Vincent. 1991. Sexual selection and the potential reproductive rates of males and females. *Nature* 351: 58–60.

Conner, J. 1988. Field measurements of natural and sexual selection in the fungus beetle, *Bolitotherus cornutus*. *Evolution* 42: 736–749.

Constantz, G. D. 1975. Behavioural ecology of mating in the male Gila topminnow, *Poeciliopsis occidentalis* (Cyprinodontiformes: Poeciliidae). *Ecology* 56: 966–973.

Constantz, G. D. 1989. Reproductive biology of poeciliid fishes. In G. K. Meffe and F. F. Snelson, eds., *Ecology and Evolution of Livebearing Fishes (Poeciliidae)*, pp. 33–50. Prentice Hall, Englewood Cliffs, N.J.

Côté, I. M., and W. Hunte. 1989. Male and female mate choice in the redlip blenny: Why bigger is better. *Anim. Behav.* 38: 78–88.

Cronly-Dillon, J., and S. C. Sharma. 1968. Effect of season and sex on the photopic spectral sensitivity of the threespined stickleback. *J. Exp. Biol.* 49: 679–687.

Crow, R. T. 1981. Behavioral adaptations to stream velocity in the guppy, *Poecilia reticulata*. M.Sc. thesis, University of British Columbia, Vancouver.

Crow, R. T., and N. R. Liley. 1979. A sexual pheromone in the guppy, *Poecilia reticulata* (Peters). *Can. J. Zool.* 57: 184–188.

Crowley, P. H., S. E. Travers, M. C. Linton, S. L. Cohn, A. Sih, and R. C. Sargent. 1991. Mate density, predation risk, and the seasonal sequence of mate choices: A dynamic game. *Am. Nat.* 137: 567–596.

Dahlgren, B. T. 1979. The effects of population density on fecundity and fertility in the guppy, *Poecilia reticulata* (Peters). *J. Fish Biol.* 15: 71–91.

Darwall, W.R.T. 1989. Sexual selection in the guppy (*Poecilia reticulata*). M.Sc. thesis. University of Utah, Salt Lake City.

Darwin, C. 1859. *On the Origin of Species by Means of Natural Selection*. John Murray, London.

Darwin, C. 1871. *The Descent of Man, and Selection in Relation to Sex*. 2d ed. John Murray, London.

Deutsch, J. C., and J. D. Reynolds. 1995. Design and sexual selection: The evolution of sex differences in mate choice. In N. S. Thompson, ed., *Perspectives In Ethology*, pp. 297–323. Plenum Press, New York.

Dominey, W. J. 1980. Female mimicry in male bluegill sunfish: A genetic polymorphism? *Nature* 284: 546–548.

Dominey, W. J. 1984. Alternative mating tactics and evolutionarily stable strategies. *Am. Zool.* 24: 385–396.

Downhower, J. F., and L. Brown. 1980. Mate preferences of female mottled sculpins, *Cottus bairdi*. *Anim. Behav.* 28: 728–734.

Dugatkin, L. 1992a. Sexual selection and imitation: Females copy the mate choice of others. *Am. Nat.* 139: 1384–1389.

Dugatkin, L. 1992b. Tendency to inspect predators predicts mortality risk in the guppy (*Poecilia reticulata*). *Behav. Ecol.* 3: 124–127.

Dugatkin, L. A. 1996. Interface between culturally based preferences and genetic preferences—female mate choice in *Poecilia reticulata*. *Proc. Natl. Acad. Sci. USA* 93: 2770–2773.

Dugatkin, L. A., and M. Alfieri. 1991a. Guppies and the tit-for-tat strategy: Preference based on past interaction. *Behav. Ecol. Sociobiol.* 28: 243–246.

Dugatkin, L. A., and M. Alfieri. 1991b. Tit-for-tat in guppies: The relative nature of cooperation and defection during predator inspection. *Evol. Ecol.* 5: 300–309.

Dugatkin, L. A., and J.-G. J. Godin. 1992a. Predator inspection, shoaling and foraging under predation hazard in the Trinidadian guppy, *Poecilia reticulata*. *Environ. Biol. Fish* 34: 265–276.

Dugatkin, L. A., and J.-G. J. Godin. 1992b. Reversal of female mate choice by copying in the guppy (*Poecilia reticulata*). *Proc. Roy. Soc. Lond. B* 249: 179–184.

Dugatkin, L. A., and J.-G.J. Godin. 1993. Female mate copying in the guppy (*Poecilia reticulata*): Age-dependent effects. *Behav. Ecol.* 4: 289–292.

Dugatkin, L. A., and R. C. Sargent. 1994. Male-male association patterns and female proximity in the guppy, *Poecilia reticulata*. *Behav. Ecol. Sociobiol.* (35): 141–145.

Dussault, G. V., and D. L. Kramer. 1981. Food and feeding behavior of the guppy, *Poecilia reticulata* (Pisces: Poeciliidae). *Can. J. Zool.* 59: 684–701.

Eberhard, W. G. 1996. Female control: Sexual selection by cryptic female choice. Princeton University Press, Princeton, N.J.

Emlen, S. T., and L. W. Oring. 1977. Ecology, sexual selection, and the evolution of mating systems. *Science* 197: 215–223.

Endler, J. A. 1977. *Geographic Variation, Speciation, and Clines*. Princeton University Press, Princeton, N.J.

Endler, J. A. 1978. A predator's view of animal color patterns. *Evol. Biol.* 11: 319–364.

Endler, J. A. 1980. Natural selection on color patterns in *Poecilia reticulata*. *Evolution* 34: 76–91.

Endler, J. A. 1983. Natural and sexual selection on color patterns in poeciliid fishes. *Environ. Biol. Fishes* 9: 173–190.

Endler, J. A. 1986. *Natural Selection in the Wild*. Princeton University Press, Princeton, N.J.

Endler, J. A. 1987. Predation, light intensity and courtship behavior in *Poecilia reticulata* (Pisces: Poeciliidae). *Anim. Behav.* 35: 1376–1385.

Endler, J. A. 1988. Sexual selection and predation risk in guppies. *Nature* 332: 593–594.

Endler, J. A. 1990. On the measurement and classification of colour in studies of animal colour patterns. *Biol. J. Linn. Soc.* 41: 315–352.

Endler, J. A. 1991. Variation in the appearance of guppy color patterns to guppies and their predators under different visual conditions. *Vision Research* 31: 587–608.

Endler, J. A. 1992. Signals, signal conditions, and the direction of evolution. *Am. Nat.* 139: S125-S153.

Endler, J. A. 1993. The color of light in forests and its implications. *Ecol. Monogr.* 63: 1–27.

Endler, J. A. 1995. Multiple-trait coevolution and environmental gradients in guppies. *TREE* 10: 22–29.

Endler, J. A., and A. E. Houde. 1995. Geographic variation in female preferences for male traits in *Poecilia reticulata. Evolution* 49: 456–468.

Endler, J. A., and T. McLellan. 1988. The processes of evolution: Towards a newer synthesis. *Ann. Rev. Ecol. Syst.* 19: 395–421.

Evans, M. R., and A. L. Thomas. 1992. The aerodynamic and mechanical effect of elongated tails in the Scarlet-Tufted Malachite Sunbird: Measuring the cost of a handicap. *Anim. Behav.* 43: 337–347.

Fajen, A., and F. Breden. 1992. Mitochondrial DNA sequence variation among natural populations of the Trinidad guppy, *Poecilia reticulata. Evolution* 46: 1457–1465.

Falconer, D. S. 1989. *Introduction to Quantitative Genetics*. Wiley, New York.

Farr, J. A. 1975. The role of predation in the evolution of social behavior of natural populations of the guppy, *Poecilia reticulata* (Pisces: Poeciliidae). *Evolution* 29: 151–158.

Farr, J. A. 1976. Social facilitation of male sexual behavior, intrasexual competition, and sexual selection in the guppy *Poecilia reticulata* (Pisces: Poeciliidae). *Evolution* 30: 707–717.

Farr, J. A. 1977. Male rarity or novelty, female choice behavior and sexual selection in the guppy *Poecilia reticulata* Peters (Pisces: Poeciliidae). *Evolution* 31: 162–168.

Farr, J. A. 1980a. The effects of sexual experience and female receptivity on courtship-rape decisions in male guppies, *Poecilia reticulata* (Pisces: Poeciliidae). *Anim. Behav.* 28: 1195–1201.

Farr, J. A. 1980b. Social behavior patterns as determinants of reproductive success in the guppy, *Poecilia reticulata* Peters (Pisces: Poeciliidae): An experimental study of the effects of intermale competition, female choice, and sexual selection. *Behaviour* 74: 38–91.

Farr, J. A. 1981. Biased sex ratios in laboratory strains of guppies, *Poecilia reticulata. Heredity* 47: 237–248.

Farr, J. A. 1983. The inheritance of quantitative fitness traits in guppies, *Poecilia reticulata* (Pisces: Poeciliidae). *Evolution* 37: 1193–1209.

Farr, J. A. 1984. Premating behavior in the subgenus *Limia* (Pisces: Poeciliidae): Sexual selection and the evolution of courtship. *Z. Tierpsychol.* 65: 152–165.

Farr, J. A. 1989. Sexual selection and secondary sexual differentiation in poeciliids: Determinants of male mating success and the evolution of female choice. In G. K. Meffe and F. F. Snelson, eds., *Ecology and Evolution of Livebearing Fishes (Poeciliidae)*, pp. 91–123. Prentice Hall, Englewood Cliffs, N.J.

Farr, J. A., and W. F. Herrnkind. 1974. A quantitative analysis of social interaction of the guppy, *Poecilia reticulata* (Pisces: Poeciliidae), as a function of population density. *Anim. Behav.* 22: 582–591.

Farr, J. A., and K. Peters. 1984. The inheritance of quantitative fitness traits in guppies, *Poecilia reticulata* (Pisces: Poeciliidae). II. Testing for inbreeding effects. *Heredity* 52: 285–296.

Farr, J. A., J. Travis, and J. C. Trexler. 1986. Behavioral allometry and interdemic variation in sexual behavior of the sailfin molly, *Poecilia latipinna* (Pisces:

Poeciliidae). *Anim. Behav.* 34: 497–509.

Fisher, R. A. 1930a. The evolution of dominance in certain polymorphic species. *Am. Nat.* 64: 385–406.

Fisher, R. A. 1930b. *The Genetic Theory of Natural Selection.* Clarendon Press, Oxford.

Fitzpatrick, S., A. Berglund, and G. Rosenqvist. 1995. Ornaments or offspring: Costs to reproductive success restrict sexual selection processes. *Biol. J. Linn. Soc.* 55: 251–260.

Folstad, I., A. M. Hope, A. Karter, and A. Skorping. 1994. Sexually selected color in male sticklebacks: A signal of both parasite exposure and parasite resistance? *Oikos* 69: 511–515.

Foo, C. L., K. R. Dinesh, T. M. Lim, W. K. Chan, and V.P.E. Phang. 1995. Inheritance of rapid markers in the guppy fish, *Poecilia reticulata. Zool. Sci.* 12: 535–541.

Forsgren, E. 1992. Predation risk affects mate choice in a gobiid fish. *Am. Nat.* 140: 1041–1049.

Forsgren, E., and C. Magnhagen. 1993. Conflicting demands in sand gobies: Predators influence reproductive behaviour. *Behaviour* 126: 125–135.

Fraser, D. F., and J. F. Gilliam. 1987. Feeding under predation hazard: Response of the guppy and Hart's rivulus from sites with contrasting predation hazard. *Behav. Ecol. Sociobiol.* 21: 203–209.

Frischknecht, M. 1993. The breeding colouration of male three-spined sticklebacks (*Gasterosteus aculeatus*) as an indicator of energy investment in vigour. *Evol. Ecol.* 7: 439–450.

Gandolfi, G. 1971. Sexual selection in relation to the social status of males in *Poecilia reticulata* (Teleostei: Poeciliidae). *Boll. Zool.* 38: 35–48.

Garcia, C. M., G. Jimenez, and B. Contreras. 1994. Correlational evidence of a sexually selected handicap. *Behav. Ecol. Sociobiol.* 35: 253–259.

Gerhardt, H. C. 1987. Evolutionary and neurobiological implications of selective phonotaxis in the gree treefrog, *Hyla cinera. Anim. Behav.* 35: 1479–1489.

Gibson, R. M., and J. W. Bradbury. 1985. Sexual selection in lekking sage grouse: Phenotypic correlates of male mating success. *Behav. Ecol. Sociobiol.* 18: 117–123.

Gibson, R. M., and J. Hoglund. 1992. Copying and sexual selection. *TREE* 7: 229–232.

Gilburn, A. S., and T. H. Day. 1994. Evolution of female choice in seaweed flies: Fisherian and good genes mechanisms operate in different populations. *Proc. Roy. Soc. Lond. B* 255: 159–165.

Gilburn, A. S., S. P. Foster, and T. H. Day. 1993. Genetic correlation between a female mating preference and the preferred male character in seaweed flies (*Coelopa frigida*). *Evolution* 47: 1788–1795.

Godin, J.-G. J. 1995. Predation risk and alternative mating tactics in male Trinidadian guppies (*Poecilia reticulata*). *Oecologia* 103: 224–229.

Godin, J.-G. J., and S. E. Briggs. 1996. Female mate choice under predation risk in the guppy. *Anim. Behav.* 51: 117–130.

Godin, J.-G. J., and S. A. Davis. 1995. Who dares, benefits: Predator approach behaviour in the guppy (*Poecilia reticulata*) deters predator pursuit. *Proc. Roy. Soc. Lond. B* 259: 193–200.

Godin, J.-G. J., and L. A. Dugatkin. 1995. Variability and repeatability of female mating preference in the guppy. *Anim. Behav.* 49: 1427–1433.

Godin, J.-G. J., and S. A. Smith. 1988. A fitness cost of the foraging in the guppy. *Nature* 333: 69–71.

Goldschmidt, T., S. A. Foster, and P. Sevenster. 1992. Inter-nest distance and sneaking in the three-spined stickleback. *Anim. Behav.* 44: 793–795.

Gong, A. 1995. Costs and benefits of female choice in the Guppy (*Poecilia reticulata*). Ph.D. diss., University of California, Los Angeles.

Gorlick, D. L. 1976. Dominance hierarchies and factors influencing dominance in the guppy *Poecilia reticulata* (Peters). *Anim. Behav.* 24: 336–346.

Grafen, A. 1990a. Sexual selection unhandicapped by the Fisher process. *J. Theor. Biol.* 144: 473–516.

Grafen, A. 1990b. Biological signals as handicaps. *J. theor. Biol.* 44: 517–546.

Grant, J.W.A., and L. D. Green. 1996. Mate copying versus preference for actively courting males by female Japanese medaka (Oryzias Latipes). *Behav. Ecol.* 7: 165–167.

Groothius, T. 1992. The influence of social experience on the development and fixation of the form of displays in the black-headed gull. *Anim. Behav.* 43: 1–14.

Gross, M. R. 1982. Sneakers, satellites and parentals: Polymorphic mating strategies in North American sunfishes. *Z. Tierpsychol.* 60: 1–26.

Gross, M. R. 1985. Disruptive selection for alternative life histories in salmon. *Nature* 313: 47–48.

Gross, M. R. 1991a. Evolution for alternative reproductive strategy: Frequency-dependent sexual selection in male bluegill sunfish. *Phil. Trans. Roy. Soc. Lond B* 332: 59–66.

Gross, M. R. 1991b. Salmon breeding behaviour and life history evolution in changing environments. *Ecology* 72: 1180–1186.

Gross, M. R., and E. L. Charnov. 1980. Alternative male life histories in bluegill sunfish. *Proc. Natl. Acad. Sci. USA* 77: 6937–6940.

Gustafsson, L., A. Qvarnstrom, and B. C. Sheldon. 1995. Trade-offs between life history traits and a secondary sexual character in male collared flycatchers. *Nature* 375: 311–313.

Gwynne, D. T. 1984. Courtship feeding increases female reproductive success in bushcrickets. *Nature* 307: 361–363.

Gwynne, D. T. 1989. Does copulation increase the risk of predation? *TREE* 4: 54–56.

Haas, R. 1976. Sexual selection in *Nothobranchius guentheri* (Pisces: Cyprinodontidae). *Evolution* 30: 614–622.

Haines, S. E., and J. L. Gould. 1994. Female platys prefer long tails. *Nature* 370: 512.

Hamilton, W. J., and M. Zuk. 1982. Heritable true fitness and bright birds: A role for parasites? *Science* 218: 384–387.

Hansen, A. J., and S. Rohwer. 1986. Coverable badges and resource defense in birds. *Anim. Behav.* 34: 69–76.

Haskins, C. P., and E. F. Haskins. 1949. The role of sexual selection as an isolating mechanism in three species of poeciliid fishes. *Evolution* 3: 160–169.

Haskins, C. P., and E. F. Haskins. 1950. Factors governing sexual selection as an isolating mechanism in the poeciliid fish *Lebistes reticulatus*. *Proc. Natl. Acad. Sci. USA* 36: 464–476.

Haskins, C. P., and E. F. Haskins. 1951. The inheritance of certain color patterns in wild populations of *Lebistes reticulatus* in Trinidad. *Evolution* 5: 216–225.

Haskins, C. P., and E. F. Haskins. 1954. Note on a "permanent" experimental alteration of genetic constitution in a natural population. *Proc. Natl. Acad. Sci. USA* 40: 627–635.

Haskins, C. P., E. F. Haskins, J.J.A. McLaughlin, and R. E. Hewitt. 1961. Polymorphism and population structure in *Lebistes reticulatus*, an ecological study. In W. F. Blair, ed., *Vertebrate Speciation*, pp. 320–395. University of Texas Press, Austin.

Haskins, C. P., P. Young, R. E. Hewitt, and E. F. Haskins. 1970. Stabilized heterozygosis of supergenes mediating certain Y-linked colour patterns in population of *Lebistes reticulatus*. *Heredity* 25: 575–589.

Hasson, O. 1989. Amplifiers and the handicap principle in sexual selection: A different emphasis. *Proc. Roy. Soc. Lond. B* 235: 383–406.

Hasson, O. 1990. The role of amplifiers in sexual selection: An integration of the amplifying and the Fisherian mechanisms. *Evol. Ecol.* 4: 277–289.

Hasson, O. 1991. Sexual displays as amplifiers: Practical examples with an emphasis on feather decorations. *Behav. Ecol.* 2: 189–197.

Hedrick, A. V. 1986. Female preferences for male calling bout duration in a field cricket. *Behav. Ecol. Sociobiol.* 19: 73–77.

Hedrick, A. V., and L. M. Dill. 1993. Mate choice by female crickets is influenced by predation risk. *Anim. Behav.* 46: 193–196.

Heinrich, W., and J. H. Schröder. 1986. Can males of *Poecilia reticulata* (Pisces: Poeciliidae) discriminate between females of different genetic quality or different residence time? *Biol. Zentrabl.* 105: 491–502.

Heisler, I. L. 1984. A quantitative genetic model for the origin of mating preferences. *Evolution* 38: 1283–1295.

Heisler, I. L. 1985. Quantitative genetic models of female choice based on "arbitrary" male characteristics. *Heredity* 55: 187–198.

Hieber, C. S., and J. A. Cohen. 1983. Sexual selection in the lovebug, *Plecia nearctica*: The role of male choice. *Evolution* 37: 987–992.

Hildeman, W. H., and E. D. Wagner. 1954. Intraspecific sperm competition in *Lebistes*. *Am. Nat.* 88: 87–91.

Hill, G. E. 1990. Female house finches prefer colourful males: Sexual selection for a condition-dependent trait. *Anim. Behav.* 40: 563–572.

Hill, G. E. 1991. Plumage coloration is a sexually selected indicator of male quality. *Nature* 350: 337–339.

Hill, G. E. 1992. Proximate basis of variation in carotenoid pigmentation in male house finches. *Auk* 109: 1–12.

Hill, G. E. 1993. Male mate choice and the evolution of female plumage coloration in the house finch. *Evolution* 47: 1515–1525.

Hill, G. E. 1994. Geographic variation in male ornamentation and female mate preference in the house finch: A comparative test of models of sexual selection. *Behav. Ecol.* 5: 64–73.

Hill, G. E., and R. Montgomerie. 1994. Plumage colour signals nutritional condition in the house finch. *Proc. Roy Soc. Lond. B* 258: 47–52.

Hill, G. E., R. Montgomerie, C. Y. Inouye, and J. Dale. 1994. Influence of dietary carotenoids on plasma and plumage colour in the house finch—intrasexual and intersexual variation. *Func. Ecol.* 8: 343–350.

Hornaday, K., S. Alexander, and F. Breden. 1995. Distribution of a repeated DNA sequence in natural populations of Trinidad guppy (*Poecilia reticulata*). *Copeia* 1995: 809–817.

Houde, A. E. 1987. Mate choice based upon naturally occurring color pattern variation in a guppy population. *Evolution* 41: 1–10.

Houde, A. E. 1988a. Genetic difference in female choice between two guppy populations. *Anim. Behav.* 36: 510–516.

Houde, A. E. 1988b. The effects of female choice and male-male competition on the mating success of male guppies. *Anim. Behav.* 36: 888–896.

Houde, A. E. 1988c. Sexual selection in guppies called into question. *Nature* 333: 711.

Houde, A. E. 1992. Sex-linked heritability of sexually selected character in a natural population of *Poecilia reticulata* (Pisces: Poeciliidae) (guppies). *Heredity* 69: 229–235.

Houde, A. E. 1993. Evolution by sexual selection: What can population comparisons tell us? *Am. Nat.* 141: 796–803.

Houde, A. E. 1994. Effect of artificial selection on male colour patterns on mating preference of female guppies. *Proc. Roy Soc. Lond. B* 256: 125–130.

Houde, A. E., and J. A. Endler. 1990. Correlated evolution of female mating preferences and male color patterns in the guppy *Poecilia reticulata*. *Science* 248: 1405–1408.

Houde, A. E., and Hankes, M. A. In press. Evolutionary mismatch of mating preferences and male color patterns in two populations of guppies (*Poecilia reticulata*). *Anim. Behav.*

Houde, A. E., and A. J. Torio. 1992. Effect of parasitic infection on male color pattern and female choice in guppies. *Behav. Ecol.* 3: 346–351.

Howard, R. D. 1988. Reproductive success in two species of anurans. In T. H. Clutton-Brock, ed., *Reproductive Success*, pp. 99–119. University of Chicago Press, Chicago.

Hubbell, S. P., and L. K. Johnson. 1987. Environmental variance in lifetime mating success, mate choice, and sexual selection. *Am. Nat.* 130: 91–112.

Hughes, A. L. 1985. Male size, mating success, and mating strategy in the mosquitofish *Gambusia affinis* (Poeciliidae). *Behav. Ecol. Sociobiol.* 17: 271–278.

Iwasa, Y., and A. N. Pomiankowski. 1994. The evolution of mate preferences for multiple sexual ornaments. *Evolution* 48: 853–867.

Iwasa, Y., and A. N. Pomiankowski. 1995. Continual change in mate preferences. *Nature* 377: 420–422.

Iwasa, Y., A. N. Pomiankowski, and S. Nee. 1991. The evolution of costly mate preferences II. The "handicap" principle. *Evolution* 45: 1431–1442.

Janetos, A. C. 1980. Strategies of female mate choice: A theoretical analysis. *Behav. Ecol. Sociobiol.* 7: 107–112.

Johnson, K., R. Thornhill, J. D. Ligon, and M. Zuk. 1993. The direction of mothers' and daughters' preferences and the heritability of male ornaments in red jungle fowl (*Gallus gallus*). *Behav. Ecol.* 4: 254–259.

Johnstone, R. A. 1995. Sexual selection, honest advertisement and the handicap principle: Reviewing the evidence. *Biol. Rev.* 70: 1–65.

Kadow, P. 1954. An analysis of sexual behaviour and reproductive physiology in the guppy, *Lebistes reticulatus* (Peters). Ph.D. diss., New York University, New York.

Kallman, K. D. 1983. The sex-determining mechanism of the poeciliid fish, *Xiphophorus montezumae*, and the genetic control of the sexual maturation process and adult size. *Copeia* 1983: 755–769.

Kallman, K. D. 1989. Genetic control of size at maturity in *Xiphorphorus*. In G. K. Meffe and F. F. Snelson, eds., *Ecology and Evolution of Livebearing Fishes (Poeciliidae)*, pp. 163–184. Prentice Hall, Englewood Cliffs, N.J.

Kallman, K. D., and V. Borkoski. 1978. A sex-linked gene controlling the onset of sexual maturity in female and male platyfish (Xiphophorus maculatus), fecundity in females and adult size in males. *Genetics* 89: 79–119.

Karino, K. 1995. Male-male competition and female mate choice through courtship display in the territorial damselfish, *Stegastes Nigricans*. *Ethology* 100: 126–138.

Karplus, I., and D. Algom. 1996. Polymorphism and pair formation in the mosquitofish *Gambusia Holbrooki* (Pisces, Poeciliidae). *Environ. Biol. Fish* 45: 169–176.

Kennedy, C.E.J. 1979. Factors influencing the sexual behaviour of the guppy *Poecilia reticulata.* Ph.D. diss., University of Leicester, U.K.

Kennedy, C.E.J., J. A. Endler, S. L. Poynton, and H. McMinn. 1987. Parasite load predicts mate choice in guppies. *Behav. Ecol. Sociobiol.* 21: 291–295.

Kirkpatrick, M., 1982. Sexual selection and the evolution of female choice. *Evolution* 36: 1–12.

Kirkpatrick, M., and L. A. Dugatkin. 1994. Sexual selection and the evolutionary effects of copying mate choice. *Behav. Ecol. Sociobiol.* 34: 443–449.

Kirkpatrick, M., and M. J. Ryan. 1991. The evolution of mating preferences and the paradox of the lek. *Nature* 350: 33–38.

Knapp, R. A., and J. T. Kovach. 1991. Courtship as an honest indicator of male parental quality in the bicolor damselfish, *Stegastes partitus*. *Behav. Ecol.* 2: 295–300.

Kodric-Brown, A. 1985. Female preference and sexual selection for male coloration in the guppy. *Behav. Ecol. Sociobiol.* 17: 199–205.

Kodric-Brown, A. 1989. Dietary carotenoids and male mating success in the guppy: An environmental component to female choice. *Behav. Ecol. Sociobiol.* 25: 393–401.

Kodric-Brown, A. 1990. Mechanisms of sexual selection: Insights from fishes. *Ann. Zool. Fennici* 27: 87–100.

Kodric-Brown, A. 1992. Male dominance can enhance mating success in guppies. *Anim. Behav.* 44: 165–167.

Kodric-Brown, A. 1993. Female choice of multiple male criteria in guppies: Interacting effects of dominance, coloration and courtship. *Behav. Ecol. Sociobiol.* 32: 415–420.

Krupa, J. J. 1995. How likely is male mate choice among anurans? *Behaviour* 132: 643–664.

Kuehne, R. A., and R. W. Barbour. 1983. *The American Darters*. University Press of Kentucky, Lexington.

Lande, R. 1981. Models of speciation by sexual selection on polygenic characters. *Proc. Natl. Acad. Sci. USA* 78: 3721–3725.

Lande, R., and M. Kirkpatrick. 1988. Ecological speciation by sexual selection. *J. Theor. Biol.* 133: 85–98.

LeBoeuf, B. J., and J. Reiter. 1988. Lifetime reproductive success in northern elephant seals. In T. H. Clutton-Brock, ed., *Reproductive Success*, pp. 344–362. University of Chicago Press, Chicago.

LeCroy, M. K. 1981. The genus Paradisaea: Display and evolution. *Am. Mus. Novit.* 2714: 1–52.

Lee, D. S., C. R. Gilbert, C. H. Hocutt, R. E. Jenkins, D. E. McAllister, and J. J. Stauffer. 1980. *Atlas of North American Fishes*. North Carolina State Museum, Raleigh.

Licht, T. 1989. Discrimination between hungry and satiated predators: The response of guppies (*Poecilia reticulata*) from high and low predation sites. *Ethology* 82: 238–243.

Liley, N. R. 1966. Ethological isolating mechanisms in four sympatric species of poeciliid fishes. *Behaviour* (Suppl.) 13: 1–197.

Liley, N. R., and B. H. Seghers. 1975. Factors affecting the morphology and behavior of guppies in Trinidad. In G. P. Baerends, C. Beer, and A. Manning, eds., *Function and Evolution in Behavior*, pp. 92–118. Oxford University Press, Oxford.

Lima, S. L., and L. M. Dill. 1990. Behavioral decisions made under the risk of predation: A review and prospectus. *Can. J. Zool.* 68: 610–640.

Long, K. D., and A. E. Houde. 1989. Orange spots as a visual cue for female mate choice in the guppy (*Poecilia reticulata*). *Ethology* 82: 316–324.

Lorenz, K. 1962. The function of colour in coral reef fishes. *Proc. Roy. Inst. Great Brit.* 39: 282–296.

Losey, G. S. J., F. G. Stanton, T. M. Telecky, and W.A.I. Tyler. 1986. Copying others, and evolutionarily stable strategy for mate choice: A model. *Am. Nat.* 128: 653–664.

Lozano, G. A. 1994. Carotenoids, parasites, and sexual selection. *Oikos* 70: 309–311.

Luyten, P. H., and N. R. Liley. 1985. Geographic variation in the sexual behaviour of the guppy, *Poecilia reticulata* (Peters). *Behaviour* 95: 164–179.

Luyten, P. H., and N. R. Liley. 1991. Sexual selection and competitive mating success of male guppies (*Poecilia reticulata*) from four Trinidad populations. *Behav. Ecol. Sociobiol.* 28: 329–336.

Lyles, A. M. 1990. Genetic variation and susceptibility to parasites: *Poecilia reticulata* infected with *Gyrodactylus turnbulli.* Ph.D. diss., Princeton University, Princeton, N.J.

Madhavi, R., and R. M. Anderson. 1985. Variability in the susceptibility of the fish host, *Poecilia reticulata*, to infection with *Gyrodactylus bullatarudis* (Monogenea). *Parasitology* 91: 531–544.

Magnhagen, C. 1991. Predation risk as a cost of reproduction. *TREE* 6: 183–186.

Magurran, A. E., and M. A. Nowak. 1991. Another battle of the sexes: The consequences of sexual asymmetry in mating cost and predation risk in the guppy, *Poecilia reticulata. Proc. Roy. Soc. Lond. B* 246: 31–38.

Magurran, A. E., and B. H. Seghers. 1990a. Population differences in predator recognition and attack cone avoidance in the guppy *Poecilia reticulata. Anim. Behav.* 40: 443–452.

Magurran, A. E., and B. H. Seghers. 1990b. Population differences in the schooling behaviour of newborn guppies, *Poecilia reticulata. Ethology* 84: 334–342.

Magurran, A. E., and B. H. Seghers. 1990c. Risk sensitive courtship in the guppy (*Poecilia reticulata*). *Behaviour* 112: 194–201.

Magurran, A. E., and B. H. Seghers. 1991. Variation in schooling and aggression amongst guppy (*Poecilia reticulata*) populations in Trinidad. *Behaviour* 118: 214–234.

Magurran, A. E., and B. H. Seghers. 1994a. A cost of sexual harassment in the guppy, *Poecilia reticulata. Proc. Roy. Soc. Lond. B* 255: 89–92.

Magurran, A. E., and B. H. Seghers. 1994b. Predator inspection behavior covaries with schooling tendency amongst wild guppy, *Poecilia reticulata*, populations in Trinidad. *Behaviour* 128: 121–134.

Magurran, A. E., and B. H. Seghers. 1994c. Sexual conflict as a consequence of ecology: Evidence from guppy, *Poecilia reticulata*, populations in Trinidad. *Proc. Roy. Soc. Lond. B* 255: 31–36.

Magurran, A. E., B. E. Seghers, G. R. Carvalho, and P. W. Shaw. 1992. Behavioural consequences of an artificial introduction of guppies (*Poecilia reticulata*) in North Trinidad: Evidence for the evolution of anti-predator behaviour in the wild. *Proc. Roy. Soc. Lond. B* 248: 117–122.

Magurran, A. E., B. H. Seghers, G. R. Carvalho, and P. W. Shaw. 1993. Evolution of adaptive variation in antipredator behaviour. *Mar. Behav. Physiol.* 23: 29–44.

Magurran, A. E., B. H. Seghers, P. W. Shaw, and G. R. Carvalho. 1995. The behavioural diversity and evolution of guppy, *Poecilia reticulata*, populations in Trinidad. *Adv. Stud. Behav.* 24: 155–202.

Marchetti, K. 1993. Dark habitats and bright birds illustrate the role of the environment in species divergence. *Nature* 362: 149–152.

Markow, T. A. 1988. *Drosophila* males provide a material contribution to offspring sired by other males. *Func. Ecol.* 2: 77–79.

Mattingly, H. T., and M. J. Butler. 1994. Laboratory predation of the Trinidadian guppy: Implications for the size-selective predation hypothesis and guppy life history evolution. *Oikos* 69: 54–64.

Maynard Smith, J. 1991. Theories of sexual selection. *TREE* 6: 146–151.

McKinnon, J. S. 1995. Video mate preferences of female three-spined sticklebacks from populations with divergent male coloration. *Anim. Behav.* 50: 1645–1655.

McLennan, D. A. 1995. Male mate choice based upon female nuptial coloration in the brook stickleback, *Culaea inconstans* (Kirtland). *Anim. Behav.* 50: 213–221.

McLennan, D. A., and J. D. McPhail. 1989a. Experimental investigations of the evolutionary significance of sexually dimorphic nuptial colouration in *Gasterosteus aculeatus* (L.): Temporal changes in the structure of the male signal. *Can. J. Zool.* 67: 1767–1777.

McLennan, D. A., and J. D. McPhail. 1989b. Experimental investigations of the evolutionary significance of sexually dimorphic nuptial colouration in *Gasterosteus aculeatus* (L.): The relationship between male colour and male behaviour. *Can. J. Zool.* 67: 1778–1782.

McMinn, H. 1990. Effects of the nematode parasite *Camallanus cotti* on sexual and non-sexual behaviors in the guppy (*Poecilia reticulata*). *Am. Zool.* 30: 245–249.

McPeek, M. A. 1992. Mechanisms of sexual selection operating on body size in the mosquitofish (*Gambusia holbrooki*). *Behav. Ecol.* 3: 1–12.

McPhail, J. D. 1969. Predation and the evolution of a stickleback (*Gasterosteus*). *J. Fish. Res. Board Can.* 26: 3183–3208.

Meffe, G. K., and F. F. Snelson. 1989a. *Ecology and Evolution of Livebearing Fishes (Poeciliidae).* Prentice Hall, Englewood Cliffs, N.J.

Meffe, G. K., and F. F. Snelson. 1989b. An ecological overview of poeciliid fishes. In G. K. Meffe and F. F. Snelson, eds., *Ecology and Evolution of Livebearing Fishes (Poeciliidae)*, pp. 13–31. Prentice Hall, Englewood Cliffs, N.J.

Meyer, A., T. D. Kocher, P. Basasibwaki, and A. C. Wilson. 1990. Monophyletic origin of Lake Victoria cichlid fishes suggested by mitochondrial DNA sequences. *Nature* 347: 550–553.

Meyer, A., J. M. Morrissey, and M. Schartl. 1994. Recurrent origin of a sexually selected trait in *Xiphophorus* fishes inferred from a molecular phylogeny. *Nature* 368: 539–542.

Meyer, J. H., and N. R. Liley. 1982. The control of production of a sexual pheromone in the female guppy *Poecilia reticulata*. *Can. J. Zool.* 60: 1505–1510.

Michod, R. E., and O. Hasson. 1990. On the evolution of reliable indicators of fitness. *Am. Nat.* 135: 788–808.

Milinski, M., and T.C.M. Bakker. 1990. Female sticklebacks use male coloration in mate choice and hence avoid parasitized males. *Nature* 344: 330–333.

Milinski, M., and T.C.M. Bakker. 1992. Costs influence sequential mate choice in sticklebacks, *Gasterosteus aculeatus*. *Proc. Roy. Soc. Lond. B* 250: 229–233.

Mitchell, S. L. 1990. The mating system genetically affects offspring performance in Woodhouse's toad (*Bufo woodhousei*). *Evolution* 44: 502–519.

Møller, A. P. 1988. Female choice selects for male sexual tail ornaments in the monogamous swallow. *Nature* 332: 640–642.

Møller, A. P. 1989. Viability costs of male tail ornaments in a swallow. *Nature* 339: 132–135.

Møller, A. P. 1990a. Effects of a haematophagous mite on the barn swallow (*Hirundo rustica*): A test of the Hamilton and Zuk hypothesis. *Evolution* 44: 771–784.

Møller, A. P. 1990b. Fluctuating asymmetry in male sexual ornaments may reliably reveal male quality. *Anim. Behav.* 40: 1185–1187.

Møller, A. P. 1992. Female swallow preference for symmetrical male sexual ornaments. *Nature* 357: 238–240.

Møller, A. P. 1994a. Repeatability of female choice in monogamous swallow. *Anim. Behav.* 47: 643–648.

Møller, A. P. 1994b. *Sexual Selection and the Barn Swallow*. Oxford University Press, Oxford.

Møller, A. P., and J. Höglund. 1991. Patterns of fluctuating asymmetry in avian feather ornaments: Implications for models of sexual selection. *Proc. Roy. Soc. Lond. B* 245: 1–5.

Møller, A. P., and A. N. Pomiankowski. 1993a. Fluctuating asymmetry and sexual selection. *Genetica* 89: 267–279.

Møller, A. P., and A. N. Pomiankowski. 1993b. Why have birds got multiple sexual ornaments? *Behav. Ecol. Sociobiol.* 32: 167–176.

Moodie, G.E.E. 1972. Predation, natural selection and adaptation in an unusual threespine stickleback. *Heredity* 28: 155–167.

Moore, A. J. 1994. Genetic evidence for the "good genes" process of sexual selection. *Behav. Ecol. Sociobiol.* 35: 235–241.

Moore, A. J., and P. J. Moore. 1988. Female strategy during mate choice: Threshold assessment. *Evolution* 42: 387–391.

Morris, M. R., M. Mussel, and M. J. Ryan. 1995. Vertical bars on male *Xiphophorus multilineatus:* A signal that deters rival males and attracts females. *Behav. Ecol.* 6: 274–279.

Munsell (Color Company). 1976. Munsell book of color, glossy finish collection. 2 vols. Munsell/Macbeth/Kollmorgan Corp. Baltimore.

Nakatsuru, K., and D. L. Kramer. 1982. Is sperm cheap? Limited male fertility and female choice in the lemon tetra (Pisces, Characidae). *Science* 216: 753–755.

Nelson, C. M., and K. Planes. 1993. Female choice of nonmelanistic males in laboratory populations of the mosquitofish, *Gambusia holbrooki. Copeia* 1993: 1143–1148.

Nichols, R. A., and R. K. Butlin. 1989. Does runaway sexual selection work in finite populations? *J. Evol. Biol.* 2: 299–313.

Nicoletto, P. F. 1991. The relationship between male ornamentation and swimming performance in the guppy, *Poecilia reticulata. Behav. Ecol. Sociobiol.* 28: 365–370.

Nicoletto, P. F. 1993. Female sexual response to condition-dependent ornaments in the guppy, *Poecilia reticulata. Anim. Behav.* 46: 441–450.

Nicoletto, P. F. 1995. Offspring quality and female choice in the guppy, *Poecilia reticulata. Anim. Behav.* 49: 377–387.

Noltie, D. B., and P. H. Johansen. 1986. Laboratory studies of microhabitat selection by the guppy, *Poecilia reticulata* (Peters). *J. Fresh. Ecol.* 3: 299–307.

Noonan, K. C. 1983. Female mate choice in the cichlid fish *Cichlasoma nigrofasciatum. Anim. Behav.* 31: 1005–1010.

Nordell, S. E. 1995. Mechanisms of selection: Mate choice and predation assessment in guppies (*Poecilia reticulata*). University of New Mexico Press, Albuquerque.

Norris, K. J. 1990a. Female choice and the evolution of the conspicuous plumage coloration of monogamous male great tits. *Behav. Ecol. Sociobiol.* 26: 129–138.

Norris, K. J. 1990b. Female choice and the quality of parental care in the great tit, *Parus major. Behav. Ecol. Sociobiol.* 27: 275–281.

Norris, K. 1993. Heritable variation in a plumage indicator of viability in male great tits *Parus major. Nature* 362: 537–539.
Owens, I.P.F., and D.B.A. Thompson. 1994. Sex differences, sex ratios, and sex roles. *Proc. Roy. Soc. Lond. B* 258: 93–99.
Parenti, L. R., and M. Rauchenberger. 1989. Systematic overview of the poeciliines. In G. K. Meffe and F. F. Snelson, eds., *Ecology and Evolution of Livebearing Fishes (Poeciliidae)*, pp. 3–12. Prentice Hall, Englewood Cliffs, N.J.
Parsons, P. A. 1990. Fluctuating asymmetry: An epigenetic measure of stress. *Biol. Rev.* 65: 131–145.
Partridge, L. 1983. Non-random mating and offspring fitness. In P. Bateson, ed., *Mate Choice*, pp. 227–255. Cambridge University Press, Cambridge, U.K.
Partridge, L., and J. A. Endler. 1987. Life history constraints on sexual selection. In J. W. Bradbury and M. B. Andersson, eds., *Sexual Selection: Testing the Alternatives*, pp. 265–277. Wiley, Chichester, U.K.
Pocklington, R., and L. M. Dill. 1995. Predation on females or males: Who pays for bright male traits? *Anim. Behav.* 49: 1122–1124.
Pomiankowski, A. N. 1987. The costs of choice in sexual selection. *J. Theor. Biol.* 128: 195–218.
Pomiankowski, A. N. 1988. The evolution of female mate preferences for male genetic quality. *Ox. Surv. Evol. Biol.* 5: 136–184.
Pomiankowski, A. N., and Y. Iwasa. 1993. Evolution of multiple sexual preferences by Fisher's runaway process of sexual selection. *Proc. Roy. Soc. Lond. B* 253: 173–181.
Pomiankowski, A., and L. Sheridan. 1994a. Linked sexiness and choosiness. *TREE* 9: 242–244.
Pomiankowski, A., and L. Sheridan. 1994b. Female choice and genetic correlations. Reply. *TREE* 9: 343.
Pomiankowski, A. N., Y. Iwasa, and S. Nee. 1991. The evolution of costly mate preferences, I. Fisher and biased mutation. *Evolution* 45: 1422–1430.
Poulin, R., and W. L. Vickery. 1993. Parasite distribution and virulence: Implications for parasite-mediated sexual selection. *Behav. Ecol. Sociobiol.* 33: 429–436.
Price, T. 1996. An association of habitat with color dimorphism in finches. *Auk* 113: 256–257.
Price, T., D. Schluter, and N. E. Heckman. 1993. Sexual selection when the female directly benefits. *Biol. J. Linn. Soc.* 48: 187–211.
Promislow, D.E.L., R. Montgomerie, and T. E. Martin. 1992. Mortality costs of sexual dimorphism in birds. *Proc. Roy. Soc. Lond. B* 250: 143–150.
Pruett-Jones, S. 1992. Independent versus nonindependent mate choice: Do females copy each other? *Am. Nat.* 140: 1000–1009.
Pruett-Jones, S. G., and M. A. Pruett-Jones. 1990. Sexual selection through female choice in Lawes' parotia, a lek-mating bird of paradise. *Evolution* 44: 486–501.
Pruett-Jones, S. G., M. A. Pruett-Jones, and H. I. Jones. 1990. Parasites and sexual selection in birds of paradise. *Am. Zool.* 30: 287–298.
Pyron, M. 1995. Mating patterns and a test for female mate choice in *Etheostoma spectabile* (Pisces, Percidae). *Behav. Ecol. Sociobiol.* 36: 407–412.
Real, L. 1990. Search theory and mate choice. I. Models of single-sex discrimina-

tion. *Am. Nat.* 136: 376–404.

Real, L. A. 1991. Search theory and mate choice. II. Mutual interaction, assortative mating, and equilibrium variation in male female fitness. *Am. Nat.* 138: 901–917.

Reimchen, T. E. 1989. Loss of nuptial color in threespine sticklebacks (*Gasterosteus aculeatus*). *Evolution* 43: 450–460.

Reynolds, J. D. 1993. Should attractive individuals court more? Theory and a test. *Am. Nat.* 141: 914–927.

Reynolds, J. D., and I. M. Côté. 1995. Direct selection on mate choice: Female redlip blennies pay more for better mates. *Behav. Ecol.* 6: 175–181.

Reynolds, J. D., and M. R. Gross. 1990. Costs and benefits of female mate choice: Is there a lek paradox? *Am. Nat.* 136: 230–243.

Reynolds, J. D., and M. R. Gross. 1992. Female mate preference enhances offspring growth and reproduction in a fish, *Poecilia reticulata*. *Proc. Roy. Soc. Lond. B* 250: 57–62.

Reynolds, J. D., M. R. Gross, and M. J. Coombs. 1993. Environmental conditions and male morphology determine alternative mating behaviour in Trinidadian guppies. *Anim. Behav.* 45: 145–152.

Reznick, D. N. 1982. Genetic determination of offspring size in the guppy (*Poecilia reticulata*). *Am. Nat.* 120: 181–188.

Reznick, D. N. 1983. The structure of guppy life histories: The tradeoff between growth and reproduction. *Ecology* 64: 862–873.

Reznick, D. N. 1989. Life history evolution in guppies. II. Repeatability of field observations and the effects of season on life histories. *Evolution* 43: 1285–1297.

Reznick, D. N., and H. Bryga. 1987. Life history evolution in guppies (*Poecilia reticulata*). I. Phenotypic and genetic changes in an introduction experiment. *Evolution* 41: 1370–1385.

Reznick, D. N., and H. A. Bryga. 1996. Life history evolution in guppies (*Poecilia reticulata*, Poeciliidae). V. Genetic basis of parallelism in life histories. *Am. Nat.* 147: 339–359.

Reznick, D. N., and J. A. Endler. 1982. The impact of predation on life history evolution in Trinidadian guppies (*Poecilia reticulata*). *Evolution* 36: 160–177.

Reznick, D. N., H. Bryga, and J. A. Endler. 1990. Experimentally induced life-history evolution in a natural population. *Nature* 346: 357–359.

Reznick, D. N., and A. P. Yang. 1993. The influence of fluctuating resources on life history: Patterns of allocation and plasticity in female guppies. *Ecology* 74: 2011–2019.

Reznick, D. N., F. H. Rodd, and M. Cardenas. 1996. Life history evolution in guppies (*Poecilia reticulata*, Poeciliidae). IV. Parallelism in life history phenotypes. *Am. Nat.* 147: 319–338.

Reznick, D. N., F. H. Rodd, F. Shaw, and R. Shaw. Submitted. An experimental evaluation of the rate of evolution in natural populations of guppies (*Poecilia reticulata*).

Riechert, S. E., and F. D. Singer. 1995. Investigation of potential male mate choice in a monogamous spider. *Anim. Behav.* 49: 715–723.

Rodd, F. H. 1994. Phenotypic plasticity in the life history traits and sexual behaviour of Trinidadian guppies (*Poecilia reticulata)* in response to their social environment. Ph.D. diss., York University, Canada.

Rodd, F. H., and D. N. Reznick. 1991. Life history evolution in guppies. III. The impact of prawn predation on guppy life histories. *Oikos* 62: 13–19.

Rodd, F. H., and M. B. Sokolowski. 1995. Complex origins of variation in the sexual behaviour of male Trinidadian guppies, *Poecilia reticulata*: Interactions between social environment, heredity, body size, and age. *Anim. Behav.* 49: 1139–1159.

Rosen, D. E., and R. M. Bailey. 1963. The poeciliid fishes (Cyprinodontiformes), their structure, zoogeography and systematics. *Bull. Am. Mus. Nat. Hist.* 126: 1–176.

Rosen, D. E., and A. Tucker. 1961. Evolution of secondary sexual characters and sexual behavior patterns in a family of viviparous fishes (Cyprinodontiformes: Poeciliidae). *Copeia* 1961: 201–212.

Rosenqvist, G. 1990. Male mate choice and female-female competition for mates in the pipefish *Nerophis ophidion*. *Anim. Behav.* 39: 1110–1115.

Rosenqvist, G., and A. E. Houde. In press. Prior exposure to male phenotypes influences mate choice in the guppy (*Poecilia reticulata*). *Behav. Ecol.*

Rosenqvist, G., and K. Johansson. 1995. Male avoidance of parasitized females explained by direct benefits in a pipefish. *Anim. Behav.* 49: 1039–1045.

Rosenthal, G. G., C. S. Evans, and W. L. Miller. 1996. Female preference for dynamic traits in the green swordtail, *Xiphophorus helleri*. *Anim. Behav.* 51: 811–820.

Rosenthal, H. L. 1951. The birth process of the guppy, *Lebistes reticulatus*. *Copeia* 1951: 304.

Rowland, W. J. 1982. The effects of male nuptial coloration on stickleback aggression: A reexamination. *Behaviour* 80: 118–126.

Rowland, W. J. 1989a. Mate choice and the supernormality effect in female stickleback (*Gasterosteus aculeatus*). *Behav. Ecol. Sociobiol.* 24: 433–438.

Rowland, W. J. 1989b. The effects of body size, aggression and nuptial coloration on competition for territories in male threespine sticklebacks, *Gasterosteus aculeatus*. *Anim. Behav.* 37: 282–289.

Rowland, W. J. 1994. Proximate determinants of stickleback behavior: An evolutionary perspective. In M. A. Bell and S. A. Foster, eds., *The Evolutionary Biology of the Threespine Stickleback*, pp. 297–344. Oxford University Press, Oxford.

Rowland, W. J. 1995. Do female sticklebacks care about male courtship vigour? Manipulation of display tempo using video playback. *Behaviour* 132: 951–961.

Rush, V. In press. The effect of thyroxin on female preferences in the guppy, *Poecilia reticulata*. *Behav. Ecol.*

Ryan, M. J. 1990. Sexual selection, sensory systems, and sensory exploitation. *Ox. Surv. Evol. Biol.* 7: 156–195.

Ryan, M. J., and B. A. Causey. 1989. "Alternative" mating behavior in the swordtail *Xiphophorus nigrensis* and *Xiphophorus pygmaeus* (Pisces: Poeciliidae). *Behav. Ecol. Sociobiol.* 24: 341–348.

Ryan, M. J., and A. Keddy-Hector. 1992. Directional patterns of female mate choice and the role of sensory biases. *Am. Nat.* 139: S4-S35.

Ryan, M. J., and A. S. Rand. 1990. The sensory basis of sexual selection for complex calls in the tungara frog, *Physalaemus pustulosus* (sexual selection for sen-

sory exploitation). *Evolution* 44: 305–314.
Ryan, M. J., and A. S. Rand. 1995. The response of female tungara frogs to ancestral mating calls. *Science* 269: 390–392.
Ryan, M. J., and W. E. Wagner. 1987. Asymmetries in mating preferences between species: Female swordtails prefer heterospecific males. *Science* 236: 595–597.
Ryan, M. J., and W. Wilczynski. 1988. Coevolution of sender and receiver: Effect on local mate preference in cricket frogs. *Science* 240: 1786–1788.
Ryan, M. J., M. D. Tuttle, and A. S. Rand. 1982. Bat predation and sexual advertisement in a neotropical anuran. *Am. Nat.* 119: 136–139.
Ryan, M. J., R. B. Coroft, and W. Wilczynski. 1990a. The role of environmental selection in intraspecific divergence of mate recognition signals in the cricket frog, *Acris crepitans*. *Evolution* 44: (1869–1872).
Ryan, M. J., J. H. Fox, W. Wilczynski, and A. S. Rand. 1990b. Sexual selection for sensory exploitation in the frog *Physalaemus pustulosus*. *Nature* 343: 66–67.
Ryan, M. J., D. K. Hews, and W.E.J. Wagner. 1990c. Sexual selection on alleles that determine body size in the swordtail *Xiphophorus nigrensis*. *Behav. Ecol. Sociobiol.* 26: 231–237.
Ryan, M. J., S. A. Perrill, and W. Wilczynski. 1992. Auditory tuning and call frequency predict population-based mating preferences in the cricket frog, *Acris crepitans*. *Am. Nat.* 139: 1370–1383.
Sargent, R. C., M. R. Gross, and E. P. Van den Berghe. 1986. Male mate choice in fishes. *Anim. Behav.* 34: 545–550.
Sato, A., F. Figueroa, C. Ohuigin, D. N. Reznick, and J. Klein. 1996. Identification of major histocompatibility complex genes in the guppy, *Poecilia reticulata*. *Immunogenetics* 43: 38–49.
Schlupp, I., C. Marler, and M. Ryan. 1994. Benefit to male sailfin mollies of mating with heterospecific females. *Science* 263: 373–374.
Schluter, D. 1988. Estimating the form of natural selection on a quantitative trait. *Evolution* 42: 849–861.
Schluter, D., and T. Price. 1993. Honesty, perception and population divergence in sexually selected traits. *Proc. Roy. Soc. Lond. B* 253: 117–122.
Schröder, J. H., and K. Peters. 1988. Differential courtship activity of competing guppy males (*Poecilia reticulata* Peters; Pisces: Poeciliidae) as an indicator for low concentrations of aquatic pollutants. *Bull. Environ. Contam. Toxicol.* 40: 396–404.
Scott, M. E., and R. M. Anderson. 1984. The population dynamics of *Gyrodactylus bullatarudis* (Monogenea) within laboratory populations of the fish host *Poecilia reticulata*. *Parasitology* 89: 159–194.
Seger, J. 1985. Unifying genetic models for the evolution of female choice. *Evolution* 39: 1185–1193.
Seghers, B. H. 1973. An analysis of geographic variation in the antipredator adaptations of the guppy, *Poecilia reticulata*. Ph.D. diss., University of British Columbia, Vancouver.
Seghers, B. H. 1974a. Geographic variation in the responses of guppies (*Poecilia reticulata*) to aerial predators, *Oecologia* 14: 93–98.
Seghers, B. H. 1974b. Schooling behavior in the guppy (*Poecilia reticulata*): An evolutionary response to predation. *Evolution* 28: 486–489.

Seghers, B. H., and A. E. Magurran. 1995. Population differences in the schooling behaviour of the Trinidad guppy, *Poecilia reticulata*: Adaptation or constraint? *Can. J. Zool.* 73: 1100–1105.

Semler, D. E. 1971. Some aspects of adaptation in a polymorphism for breeding colours in the threespine stickleback (*Gasterosteus aculeatus*). *J. Zool.* 165: 291–302.

Shaw, P. W., G. R. Carvalho, A. E. Magurran, and B. H. Seghers. 1991. Population differentiation in Trinidadian guppies (*Poecilia reticulata*): Patterns and problems. *J. Fish Biol.* 39 (Supp. A): 203–209.

Shaw, P. W., G. R. Carvalho, B. H. Seghers, and A. E. Magurran. 1992. Genetic consequences of an artificial introduction of guppies (*Poecilia reticulata*) in N. Trinidad. *Proc. Roy. Soc. Lond. B* 248: 111–116.

Shaw, P. W., G. R. Carvalho, A. E. Magurran, and B. H. Seghers. 1994. Factors affecting the distribution of genetic variability in the guppy, *Poecilia reticulata. J. Fish Biol.* 45: 875–888.

Shelly, T. E., and W. J. Bailey. 1992. Experimental manipulation of mate choice by male katydids: The effect of female encounter rate. *Behav. Ecol. Sociobiol.* 30: 277–282.

Sih, A. 1988. The effects of predators on habitat use, activity and mating behaviour in a semi-aquatic bug. *Anim. Behav.* 36: 1846–1848.

Sih, A. 1994. Predation risk and the evolutionary ecology of reproductive behaviour. *J. Fish Biol.* 45: 111–130.

Sih, A., J. Krupa, and S. Travers. 1990. An experimental study on the effects of predation risk and feeding regime on the mating behaviour of the water strider. *Am. Nat.* 135: 284–290.

Simmons, L. W., T. Llorens, M. Schinzig, D. Hosken, and M. Craig. 1994. Sperm competition selects for male mate choice and protandry in the bushcricket, *Requena verticalis* (Orthoptera, Tettigoniidae). *Anim. Behav.* 47: 117–122.

Smith, H. G., and R. Montgomerie. 1991. Sexual selection and the tail ornaments of North American barn swallows. *Behav. Ecol. Sociobiol.* 28: 195–201.

Snelson, F. F. 1989. Social and environmental control of life history traits in poeciliid fishes. In G. K. Meffe and F. F. Snelson, eds., *Ecology and Evolution of Livebearing Fishes (Poeciliidae)*, pp. 149–161. Prentice Hall, Englewood Cliffs, N.J.

Spieser, O. H., and J. H. Schröder. 1980. Differential response to irradiation in offspring of freshwater and seawater substrains of *Poecilia (Lebistes) reticulata* Peters in the "guppy male courtship activity test." *Theor. Appl. Genet.* 51: 223–232.

Stearns, S. C. 1992. *The Evolution of Life Histories.* Oxford University Press, Oxford.

Stoner, G., and F. Breden. 1988. Phenotypic differentiation in female preference related to geographic variation in male predation risk in the Trinidad guppy (*Poecilia reticulata*). *Behav. Ecol. Sociobiol.* 22: 285–291.

Strauss, R. E. 1990. Predation and life history variation in *Poecilia reticulata* (Cyprinodontiformes: Poeciliidae). *Environ. Biol. Fish* 27: 121–130.

Sumner, I. T., J. Travis, and C. D. Johnson. 1994. Methods of female fertility advertisement and variation among males in responsiveness in the sailfin molly (*Poecilia latipinna*). *Copeia* 1994: 27–34.

Swaddle, J. P., and I. C. Cuthill. 1994a. Female zebra finches prefer males with symmetric chest plumage. *Proc. Roy. Soc. Lond. B* 258: 267–271.

Swaddle, J. P., and I. C. Cuthill. 1994b. Preference for symmetric males by female zebra finches. *Nature* 367: 165–166.

Thornhill, R. 1979. Male and female sexual selection and the evolution of mating systems in insects. In M. S. Blum and N. A. Blum, eds., *Sexual Selection and Reproductive Competition in Insects*, pp. 81–121. Academic Press, New York.

Thornhill, R. 1980. Rape in *Panorpa* scorpionflies and a general rape hypothesis. *Anim. Behav.* 28: 52–59.

Thornhill, R. 1981. *Panorpa* (Mecoptera: Panorpidae) scorpionflies: Systems for understanding resource-defense polygyny and alternative male reproductive efforts. *Ann. Rev. Ecol. Syst.* 12: 355–386.

Thornhill, R. 1992. Female preference for the pheromone of males with low fluctuating asymmetry in the Japanese scorpionfly (*Panorpa japonica* : Mecoptera). *Behav. Ecol.* 3: 277–283.

Thornhill, R., and K. P. Sauer. 1992. Genetic sire effects on the fighting ability of sons and daughters and mating success of sons in the scorpionfly (*Panorpa vulgaris*). *Anim. Behav.* 43: 255–264.

Trail, P. W., and E. Adams. 1989. Active mate choice at cock-of-the-rock leks: Tactics of sampling and comparison. *Behav. Ecol. Sociobiol.* 25: 283–292.

Travers, S. E., and A. Sih. 1991. The influence of starvation and predators on the mating behavior of a semi-aquatic insect. *Ecology* 72: 2123–2136.

Travis, J. 1994. Size-dependent behavioural variation and genetic control within and among populations. In C.R.B. Boake, ed., *Quantitative Genetic Approaches to Animal Behavior*, pp. 165–187. University of Chicago Press, Chicago.

Travis, J., and J. C. Trexler. 1990. Phenotypic plasticity in the sailfin molly (Pisces: Poeciliidae). II. Laboratory experiment. *Evolution* 44: 157–167.

Travis, J., and B. D. Woodward. 1989. Social context and courtship flexibility in male sailfin mollies, *Poecilia latipinna* (Pisces: Poeciliidae). *Anim. Behav.* 38: 1001–1011.

Trexler, J. C., and J. Travis. 1990. Phenotypic plasticity in the sailfin molly (Pisces: Poeciliidae). I. Field experiments. *Evolution* 44: 143–156.

Tuttle, M. D., and M. J. Ryan. 1982. The role of synchronized calling, ambient light, and ambient noise in an anti-bat-predator behavior of a treefrog. *Behav. Ecol. Sociobiol.* 11: 125–131.

Verrell, P. A. 1994. Males choose larger females as mates in the salamander, *Desmognathus fuscus*. *Ethology* 99: 162–171.

Vrijenhoek, R. C. 1989. Genotypic diversity and coexistence among sexual and clonal lineages of *Poeciliopsis*. In D. Otte and J. A. Endler, eds., *Speciation and Its Consequences*, pp. 386–400. Sinauer, Sunderland, Mass.

Wade, M. J., and S. G. Pruett-Jones. 1990. Female copying increases the variance in male mating success. *Proc. Natl. Acad. Sci. USA* 87: 5749–5753.

Warner, R. R., and E. T. Schultz. 1992. Sexual selection and male characteristics in the bluehead wrasse, *Thalassoma bifasciatum*: Mating site acquisition, mating site defense and female choice. *Evolution* 46: 1421–1442.

Watson, P. J., and R. Thornhill. 1994. Fluctuating asymmetry and sexual selection. *TREE* 9: 21–25.

Werner, M., and J. H. Schröder. 1980. Mutational changes in the courtship activity of male guppies (*Poecilia reticulata*) after X-irradiation. *Behavior Genetics* 10: 427–430.

West-Eberhard, M. J. 1983. Sexual selection, social competition, and speciation. *Q. Rev. Biol.* 58: 155–183.

Wiegmann, D. D., L. A. Real, T. A. Capone, and S. Ellner. 1996. Some distinguishing features of models of search behavior and mate choice. *Am. Nat.* 147: 188–204.

Wiernasz, D. C. 1995. Male choice on the basis of female melanin pattern in *pieris* butterflies. *Anim. Behav.* 49: 45–51.

Wilczynski, W., A. S. Rand, and M. J. Ryan. 1995. The processing of spectral cues by the call analysis of the tungara frog, *Physalaemus pustulosus. Anim. Behav.* 49: 911–929.

Wilkinson, G., and P. Reillo. 1994. Female choice response to artificial selection on an exaggerated male trait in a stalk-eyed fly. *Proc. Roy. Soc. Lond. B* 255: 1–6.

Winemiller, K. O., M. Leslie, and R. Roche. 1990. Phenotypic variation in male guppies from natural inland populations: An additional test of Haskins' sexual selection/predation hypothesis. *Environ. Biol. Fish* 29: 179–191.

Winge, O. 1922a. A peculiar mode of inheritance and its cytological explanation. *J. Genetics* 12: 137–144.

Winge, O. 1922b. One-sided masculine and sex-linked inheritance in *Lebistes reticulatus. J. Genetics* 12: 145–162.

Winge, O. 1927. The location of eighteen genes in *Lebistes reticulatus. J. Genetics* 18: 1–43.

Winge, O. 1937. Succession of broods in *Lebistes. Nature* 140: 467.

Winge, O., and E. Ditlevsen. 1947. Colour inheritance and sex determination in *Lebistes. Heredity* 1: 65–83.

Woodhead, A. D., and N. Armstrong. 1985. Aspects of the mating behavior of male mollies (*Poecilia* spp.). *J. Fish Biol.* 27: 593–601.

Yasukawa, K. 1981. Male quality and female choice of mate in the red-winged blackbird (*Agelaius phoeniceus*). *Ecology* 62: 922–929.

Zahavi, A. 1975. Mate selection—a selection for a handicap. *J. Theor. Biol.* 53: 205–214.

Zimmerer, E. J., and K. D. Kallman. 1989. Genetic basis for alternative reproductive tactics in the pygmy swordtail, *Xiphophorus nigrensis. Evolution* 43: 1298–1307.

Zuk, M. 1988. Parasite load, body size, and age of wild-caught male field crickets (Orthoptera: Gryllidae): Effects on sexual selection. *Evolution* 42: 969–976.

Zuk, M. 1992. The role of parasites in sexual selection: Current evidence and future directions. *Adv. Study Behav.* 21: 39–68.

Zuk, M., and J. G. Decruyenaere. 1994. Measuring individual variation in colour: A comparison of two techniques. *Biol. J. Linn. Soc.* 53: 165–173.

Zuk, M., K. Johnson, R. Thornhill, and J. D. Ligon. 1990a. Mechanisms of female choice in the red jungle fowl. *Evolution* 44: 477–485.

Zuk, M., R. Thornhill, J. D. Ligon, K. Johnson, S. Austad, S. H. Ligon, N. W. Thornhill, and C. Costin. 1990b. The role of male ornaments and courtship behavior in female mate choice of red jungle fowl. *Am. Nat.* 136: 459–473.

Zuk, M., R. Thornhill, J. D. Ligon, and K. Johnson. 1990c. Parasites and mate choice in red jungle fowl. *Am. Zool.* 30: 235–244.

Zuk, M., T. S. Johnsen, and T. Maclarty. 1995a. Endocrine-immune interactions, ornaments and mate choice in red jungle fowl. *Proc. Roy. Soc. Lond. B* 260: 205–210.

Zuk, M., S. L. Popma, and T. S. Johnsen. 1995b. Male courtship displays, ornaments and female mate choice in captive red jungle fowl. *Behaviour* 132: 821–836.

Author Index

Subject Index

Taxonomic Index